Contents

Acknowledgements	4
Introduction	6

Rural Water Supplies

Underground water explained	10
Hand-dug wells	20
Building an upgraded shallow well	23
Hand-drilled tubewells	46
Sanitary surveys of wells and tubewells	58
Handpumps	61
The Bucket Pump	69
The Blair Pump	109
The Nsimbi Pump	142
The Bush Pump	153
Supplying water by gravity	210
The protection of naturally occurring springs	212
The siphon well	221
The gravity well	223
Rainwater harvesting	225
Water point design	235
Hygiene aspect of water supply	244
Drinking water quality for rural areas	248
The purification of water	255
Sand filtration of water	256
Water treatment with chlorine	266

Rural Sanitation

The Blair Latrine and how it works	270
Upgrading the ordinary pit latrine	287
How to build a Blair Latrine	291
The single compartment version	291
The double compartment version	312
The multicompartment version	321
A urinal for school latrines	329
The tank and soakaway version	333
Adding a flush toilet	349
Blair Latrine training programmes	351
Index	354

Acknowledgements

The technical information written in this manual is drawn from the accumulation of many years' experience, not only by members of the Blair Research Laboratory, but by many others working in the Ministry of Health at Provincial, District, Ward and Village level. We in the laboratory are much dependent on the outcome of operations in the field, where the techniques described in this book actually work in practice. I wish to acknowledge the Health Inspectorate and their staff for the full support and co-operation they have extended to my staff and me during our many years of work together in this field. We have become much wiser from having listened to those who really understand the way of life in the rural areas. Whether the feedback from field operations has been discouraging or encouraging, the effect is to improve the state of the art — and that is our aim.

The work described in this volume may never have developed in the way it has without the dedication of many people. In particular I wish to mention the enthusiasm and support of my friend and colleague, Ephraim Chimbunde who has worked alongside me for over fifteen years and has been a central figure in our field operations for all that time. I am indebted to him for his unique contribution.

Our Blair Research field team has been active for fifteen years, researching and trying new ideas and teaching all over the country. In particular the great efforts of Fambi Gono, Philimon Ndororo, Joshua Mazanza and their supporting teams should be mentioned for their consistent effort over the years. In more recent years our laboratory team, Felix Chawira and Michael Jere, have contributed much by their bacteriological testing programme.

A very large number of people have made important contributions to this work, either by direct assistance or by encouragement and support. The Chief Health Inspector, Mr John Mvududu and the National Decade Officer, Mr Naison Mtakwa, supported by Mr Bob Boydell and Mr Piers Cross of the World Bank, have given considerable assistance and support. I also extend my gratitude to the Provincial Health Inspectorate and their staff, particularly, Mr Mark Chibanda, Mr Steven Maphosa, Mr Talkmore Masaka, Mr Masotcha Mtshena, Mr Shadreck Musingarabwe, Mr Garth Parsons, Mr David Proudfoot, Mr Andrew Ruwende and Mr Nathaniel Tembo. The contributions made by Mr Ben De Beer, Mr Ernest Berk, Ms Colleen Butcher, Mr Johnston Chinyanga, Mr Ron Evans, Mr Roland Haebler, Mrs Mahezent Hapte-Mariam, Mr Jim Holland, Mr Roger Ing, Mr Norman Johnston, Mr Godwin Kawadza, Dr Richard Laing, Mr Simon Metcalf, Mr Jack Naude, Dr Alan Pugh, Mr Julian Sturgeon, Mr Erwin Von Elling, Sr Patricia Walsh, Mr Bill White and Mr David Williams are greatly appreciated. Many more could equally be added to this list — and to all I extend my appreciation.

Rural   es
and

Peter Mor

Blair Res
Ministry
Harare

MACMILLAN
PUBLISHERS

© Ministry of Health, Zimbabwe 1990

Any parts of this book, including the illustrations, may be copied, reproduced, or adapted to meet local needs, without permission from the author or publisher, provided the parts reproduced are distributed free or at cost — not for profit. For any reproduction with commercial ends, permission must first be obtained from the publisher and the author. The publisher and author would appreciate being sent a copy of any materials in which text or illustrations have been used.

First published 1990

Published by *Macmillan Publishers Ltd*
London and Basingstoke
Associated companies and representatives in Accra, Auckland, Delhi, Dublin, Gabarone, Hamburg, Harare, Hongkong, Kuala Lumpur, Lagos, Manzini, Melbourne, Mexico City, Nairobi, New York, Singapore, Tokyo

ISBN 0-333-48569-6

Printed in Hong Kong

A CIP catalogue record for this book is available from the British Library.

The publishers would like to extend their thanks to Ms Mary Cole (B. Sc (Hons) PGCE) for her editorial input on the early drafts of this book.

The author and publishers wish to thank the following who have kindly granted permission for the use of copyright material:

The Central African Journal of Medicine for the article 'The Control of Flies', Vol.23, No.1, January 1977.

The Deputy Director (Operations) of the District Development Fund, Harare, Zimbabwe, for drawings from educational material.

Every effort has been made to trace all the copyright holders, but if any have been inadvertently overlooked the publishers will be pleased to make the necessary arrangement at the first opportunity.

Many of the drawings in this work are taken from a series of field manuals developed by Mrs Sue Laver, and illustrated by Mr Kors de Waard and Mrs Colleen Cousins, to whom I am indebted. Mr de Waards' fine work is seen in the sanitation and well sections and the Blair and Bush Pumps, and Mrs Cousins' work is seen in the illustrations of the Vonder Rig. In addition I wish to thank Mrs Tali Bradley for her fine illustrations of the Bucket Pump. A few illustrations are taken from other works and these are acknowledged individually. The less artistic drawings are mostly my own.

Many non-Governmental organisations have also played an important part in our work and the preparation of written material. These include the Dominican Sisters of Zimbabwe, GTZ, Interconsult A/S Harare, The International Reference Centre for Wastes Disposal, Switzerland, Norad, Redd Barna, Save the Children's Fund UK, SIDA, UNICEF and WaterAid UK. The author and publishers extend particular thanks to SIDA for financial support in the preparation of this book.

Finally I wish to acknowledge the support and encouragement of both the Director, Blair Research Laboratory, and the Permanent Secretary for Health. Their support means much to those of us working in the field of Research and Development.

Peter Morgan

Blair Research Laboratory,
Harare

October 1989

Introduction

It is the aim of Government to help provide one Blair Latrine for every family and distribute water points so that each family has easy access to a protected supply. This amounts to a total of approximately 750,000 latrines and 75,000 protected water points to be constructed before the end of the century. This is an immense task and will require a considerable amount of hard work and dedication by all those involved.

The model of Blair Latrine varies from one place to another. Most built in the rural areas are either single or double compartment units made of brick. A multicompartment version has also been designed for schools and the design of a desludgeable unit, which can be upgraded to contain a flush system, is also available.

The design of protected water points varies considerably. Many are commercially drilled boreholes fitted with a handpump, whilst others will be hand-dug wells or hand-drilled tubewells fitted with a handpump or bucket and windlass. The emphasis is placed on tapping the huge groundwater supply and raising water to the surface. Several other systems will also be used where they are suitable, including the protection of springs and the construction of rainwater catchment systems and dams.

The handpump of choice for most installations is the time-tested Zimbabwe 'Bush Pump', which is ideally suited for deep or heavy duty use. Where village level maintenance is required, most conventional reciprocating handpumps may be too complex for villagers to repair themselves, and the Zimbabwe 'Bucket Pump' may be more suitable, since it can be managed at village level with relatively little support from outside. Where light duty pumps are required for small communities, and a system of management is operational, PVC-bodied pumps like the Blair and Nsimbi have been used with success. The ideal distribution for the village pump is at 'clan level' where greater care and a willingness to maintain are generally present. More complex Bush Pumps are maintained by Pump Minders or Pump Fitters employed by the DDF with day to day tightening of nuts and cleaning being performed by a Pump Caretaker.

The physical and bacteriological quality of water extracted from unprotected wells and water holes can be improved significantly by upgrading the headworks of the well, improving the lining and fitting a windlass. Thus family wells can be improved without the expense of fitting a handpump in the first instance. When further upgrading is required the Blair Pump may be ideally suited for a shallow family well. Family water facilities deserve much encouragement since they will always receive care and attention and can be developed alongside a vegetable garden which also has a beneficial effect on family health. In family situations, maintenance is accepted as being the responsibility of the owner/user.

This book describes constructional methods for the Blair Latrine, improved wells and springs, the fitting and maintenance of handpumps and many other aspects of rural water supply.

Many of the Bulletins presented in this manual have been simplified and presented in the form of manuals for field workers. This is a process which is continuing. Currently field manuals are available for the single and double compartment Blair Latrine, the Upgraded Well, the Bucket and Blair Pumps, and the VonderRig. Several other manuals are in the process of being prepared.

The state of the art in this sector is developing fast, and this volume attempts to describe the latest techniques being used in Zimbabwe. It is possible that some may have application outside Zimbabwe, but care should always be taken when introducing them. As a rule of thumb, it is generally far better to take an existing technology, well established in a particular country, and build on that and develop it, rather than introduce something foreign. All the major technologies used in Zimbabwe's rural water and sanitation sector have been developed in Zimbabwe itself, and this is seen as a big advantage. It is hoped that local designers in other lands will also meet this challenge and develop and adapt their own indigenous techniques for future use.

Attempts have been made throughout this volume to describe principles of design and basic concepts, which future designers can adapt for use in other lands. Many of the pumps, for instance, can be built using a variety of materials which suit local conditions. This also holds true for the sanitation technology. Ventilated pit latrines can be successfully built with a very wide range of materials.

Some sections of these Bulletins are more detailed than others. Constructional notes for the Blair Latrine, Upgraded wells and tubewells, and the installation of the Bucket and Bush Pumps are described in detail because they form the main components of Zimbabwe's Rural Water and Sanitation Programme.

Whilst this volume deals almost exclusively with technological options, it is well understood that few programmes of rural development can succeed and be sustained without a parallel development of human resources. All successful implementaion schemes must have at least two complementary components — sound technology and well-motivated participants. This volume only touches on the latter, but it is well known that one cannot work without the other.

The wise designer attempts to draw on experience which may extend over many decades, and uses techniques which are simple in concept and design. The technique of updating and adapting well-established traditional techniques rarely fails, simply because such techniques have already undergone the most severe test of all — time. A careful use of the forces found in the Natural World adds considerable strength to any design, by imparting that most valuable of properties — dependability.

PRM

Harare

December 1988

RURAL WATER SUPPLIES

The village pump — for many a vital necessity.

Underground water explained

Water fit for drinking exists in the ground in some form at some depth nearly everywhere on earth. Even the Sahara desert is underlain by water, an estimated 600,000 cubic kilometres of it, spread over 6.5 million square kilometres of land area. Almost all of the world's stock of fresh water, 8.2 million cubic kilometres, or more than 97% of the total available, is inside the earth itself and exists as groundwater. The rest exists in lakes, streams and rivers.

Most underground water is constantly in motion — being pulled by gravity from higher aquifers to lower, where it may appear in springs and streams. It is lifted from the ground by plants and by man himself. Gravity attracts water from the skies, pulls it beneath the surface of the ground, distributes it among permeable layers of the earth and influences the direction in which it flows. Rainwater seeps downwards through the soils until it is blocked by rock or non-porous layers such as clay. Simultaneously it spreads out horizontally so that vast volumes of the earth become saturated with water. The water soaks into the soil and moves through the permeable earth from pore to pore.

The proportion of water that sinks into the ground varies with the character of the soil. If the soil is dry and porous, large amounts will seep in. If a sudden downpour falls on a sloping surface of less permeable material, such as clay, most of the rainwater will run off and be lost to the soil. The outermost layers of the earth are composed largely of porous material, sands, gravels, silts and decaying vegetation. Most of this surface is underlain by porous or decomposing rocks. Beneath this everywhere is bedrock, so compact, as a result of its molten origin or of

Hydrological cycles

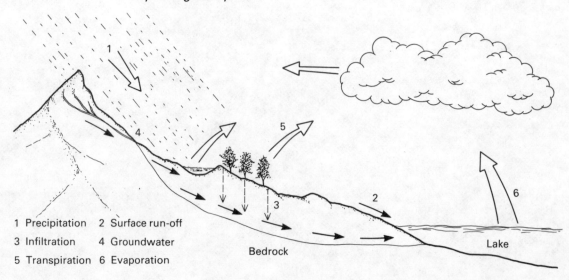

1 Precipitation 2 Surface run-off
3 Infiltration   4 Groundwater
5 Transpiration 6 Evaporation

subsequent heat and pressure, it is totally impermeable. All layers are classified by water content — the upper zone of aeration and the lower zone of saturation.

Seeping below the surface, water first enters the zone of aeration, where the soil contains both water and air. Its depth varies widely from a few centimetres, in a vlei with a high water table, to hundreds of metres in areas with a low water table. In this zone of aeration water shows its powers of adhesion by clinging to particles of soil and rock. Some water that enters this region sinks to the layers beneath, some is absorbed by plants and some evaporates into the air again.

The lower moist layer, the zone of saturation, is the earth's main reservoir. Wells and boreholes dip into it and springs, rivers and lakes are its natural outcroppings on the surface. The top of the saturated zone, the boundary between the water layer (known as the water table) and the zone of aeration, is a narrow zone called the capillary fringe where water rises as a result of capillary action into the zone immediately above the water table. The sparkle of water at the bottom of a well is an exposed part of the water table. Around it and continuous with it, the same water table extends in the ground to be exposed again in the next well or in a natural feature like a river or lake.

The groundwater itself occurs in pores, voids or fissures in the ground material. Pores are the spaces between the mineral grains in sedimentary ground layers and in decomposing rock formations. The word 'aquifer' is used to denote a layer of water bearing material.

The word aquifer comes from the two latin words, *aqua* (water) and *ferre* (to carry). The aquifer literally carries water — underground. The aquifer may be a layer of gravel or sand, a zone between lava flows or even a large body of massive rocks, such as fractured granite with large opening in it. An aquifer may only be a few metres thick or tens of hundreds of metres thick, it may lie just under the surface or hundreds of metres down, it may underlie a few hectares of land or thousands of square kilometres.

Water distribution above and within a porous aquifer.

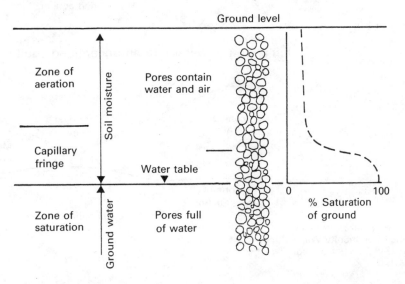

The quantity of water that a given piece of saturated earth can contain depends on the earth's *POROSITY*, the total measure of the spaces among the grains or in the cracks that can fill with water. If the grains of sand or gravel are all about the same size or 'well sorted', the spaces between them account for a large proportion of the total volume of the aquifer. If the grains however are poorly sorted, the spaces between the grains may be filled with smaller particles instead of water. Poorly sorted rocks, therefore, do not hold as much water as well sorted rocks.

If water is to move through rock or other material, the pores must be connected to one another. If the rock has a great many connecting pore spaces big enough to enable water to move freely through them, the rock is said to be permeable. A rock or underground material that is a good source of water must contain many inter-connecting pore spaces or cracks. A compact rock almost without pore spaces, such as granite, may be permeable if it contains enough sizeable cracks or fractures. The *PERMEABILITY* of the ground is a measure of the freedom with which water moves through it.

In Zimbabwe, fractures or decomposing granite can yield a great deal of water. Nearly all consolidated rock formations are broken by parallel systems of cracks called joints. At first the joints are like hairline cracks, but they tend to enlarge. Water enters the joint and gradually dissolves the rock so that it becomes decomposed. Decomposed rocks are permeable. The action of the water in decomposing the rocks

Infiltration of water into an unconfined aquifer during the wet season.

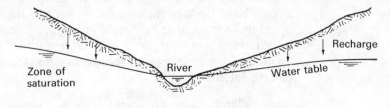

Infiltration of water into an unconfined aquifer during the dry season.

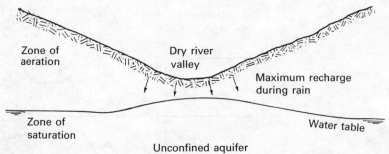

Drawings taken from
Small Community Water Supplies
WHO-IRC Technical Paper No. 18,
1981.

extends as time passes. Also the movement of water may flush out decomposed or 'weathered' rocks thereby enlarging the openings. High-yielding boreholes are those that pass through layers of decomposing granite with a high permeability.

All openings in rocks such as joints, cleavage planes and random cracks are called fissures. Igneous rocks, that is those originating from volcanic erupted material, are not generally porous unless they are decomposed by weathering. Fissures may also occur in sedimentary rocks, which are sediments compacted by glacial pressure to form solid material.

An unconfined aquifer is open to infiltration of water directly from the ground surface. Most shallow wells are dug in such situations and are common in most parts of Zimbabwe.

A confined aquifer is one where the water-bearing ground formation is capped by an impermeable ground layer. Water occurs under pressure in a confined aquifer and once a borehole is drilled water may rise up through the hole under pressure. The occurrence is not uncommon in Zimbabwe. Often a borehole will be drilled to a depth of 50 metres, and once a critical layer of water is met, the water rises, sometimes to within a few metres of the surface. If water comes to the surface under

Confined aquifer fed from a recharge area.

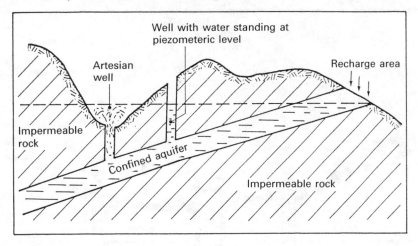

Perched water table.

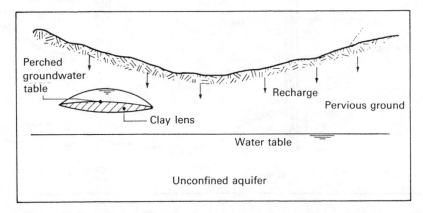

pressure it is known as a free-flowing or 'artesian well'. Several of these have been drilled in the Zambezi Valley and elsewhere.

A curiosity which exists in some places is the 'perched water table'. The infiltration of water through permeable ground from the surface may be halted where a 'lens' of impermeable material like clay is formed well above the real groundwater table.

A relationship does not necessarily exist between the water-bearing capacity of rocks and the depths at which they are found. A dense granite with a few cracks may be found at the surface, whereas a porous material may lie hundreds of metres underground. On average however, porosity and permeability decrease as depth increases. The pores and cracks in the rocks at great depth are virtually closed because of the great pressure of the overlaying rocks.

Normally hand-dug wells do not penetrate very deeply into the water-bearing strata, and can dry out in the drier months of the year. Hand-drilled tubewells penetrate deeper in water-bearing aquifers and are able to extract water from a greater vertical depth of the 'overburden', the softer layer of material that lies above the hard bedrock. Deeper boreholes penetrate right through the overburden, and tap water from decomposing rocks deep down.

Methods of tapping groundwater.

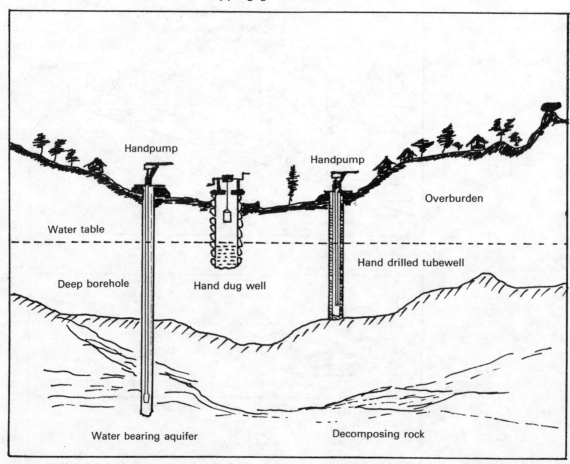

The relationship between water table and rainfall

The effect of rainfall on the underground water table is very considerable and this has been studied in the Epworth area near to Harare by Blair Research Laboratory. Fig. 1 shows the relationship between rainfall and water table during the period from November 1984 to May 1986. Fig. 2 shows a similar relationship in 1987/88 in Epworth. The graphs reveal a close relationship between rainfall and water table.

During the dry season, the water table falls slowly and consistently from the overburden. Water is lost into low-lying streams and into deeper parts of the aquifer. The water table continues to fall even after the first rains have begun. During this phase the soils of the overburden become saturated. In Fig. 1 this process was completed by December 26th 1985, when the water table stopped falling and began to rise. At this time 140 mm of rain had fallen.

Figure 1. Graph showing water table and rainfall at the Chisungu Station, Epworth. Data: Blair Research Laboratory. 1985/1986.

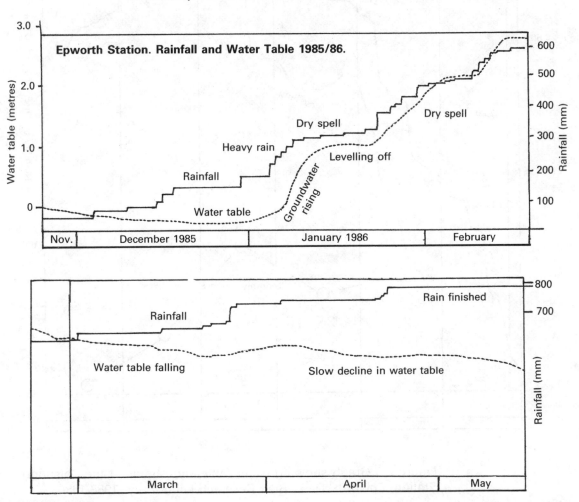

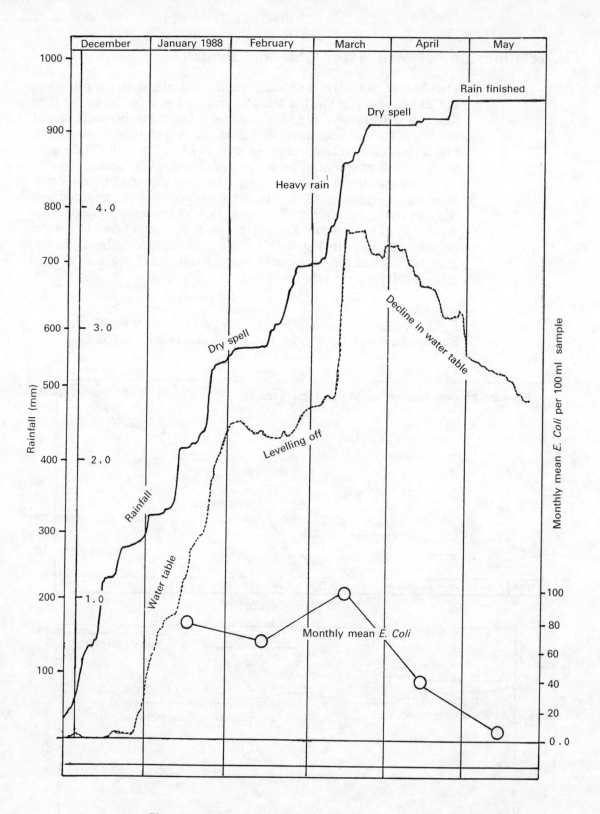

Figure 2. Graph showing water table and rainfall at the Chisungu Station, Epworth. Data: Blair Research Laboratory. 1987/1988.

After this initial period, each heavy rain was followed by a noticeable rise in the water table, the heavier the rain, the faster the rise of the water table. Of particular interest in Fig. 1 is the marked rise of groundwater between January 6th and 13th 1988 following very heavy rains and flooding.

Rapid rises in groundwater are also associated with increased flows of ground water in the horizontal direction, since water flows from higher ground to lower ground all the time. There is a constant flow of water downwards, draining from higher levels to lower levels in the terrain. In Fig. 1 the heavy rains were followed by a dry spell after January 8th 1986 and by January 14th the water table had levelled off. The water table changed little until the next spell of rains which fell between January 21st and 30th 1986. Once again the water table rose dramatically following the heavy rains. A further step in both rain and water table was recorded between February 9th and 14th 1986. After this date rainfall was much reduced, and a slow but steady decline of water table began. By February 16th 1986 a total of 600 mm of rain had fallen on the station, with a resulting rise of water table of 3200 mm. The fall in water table between February 18th and May 18th 1986 was 700 mm. The main rains ceased in late April that year. Similar movements in water can be seen on the 1987/88 graph.

This filling and draining of the upper aquifers in Epworth can be compared to filling a bath full of sand with water. In this case the bath plug has been removed and there is a constant loss of water from the sand through the plug hole. Water levels in the sand are recharged when water is added from above (the rains), but the loss is constant. Water rises higher in the sand if more water is added (after years of heavy rain), and is lowered if less water is added (after drier years). The aquifer is so large that the artificial removal of water through handpumps has an insignificant effect on the source. However, in other situations, where more powerful motorised pumps are used to extract water for irrigation from an aquifer, the source may be depleted considerably.

Relationship between rate of rise of water table and water quality

An interesting relationship also exists between the heavy rains and the quality of the groundwater. The rate of flow of water through the ground is higher following heavy rains compared with periods of lighter rain and no rain. Since the water in the aquifer is 'topped up' more rapidly after heavy rains, and both the horizontal and vertical migrations of water are accelerated at such times, it is not surprising that bacteria held in the water are also carried through greater distances after periods of heavy rain. Bacteria on the surface are also carried more deeply into the overburden after heavy rains. Generally the closer the groundwater is to the surface, the more influential is the effect of heavy rain in carrying bacteria and other organisms through the soil.

Pathogenic bacteria do not find the soil a suitable medium in which to survive, and die off as they pass through it. Since water acts as a vehicle to carry them, reduced water movement restricts their passage through the soil and greatly reduces their ability to contaminate vital sources of water before they die off.

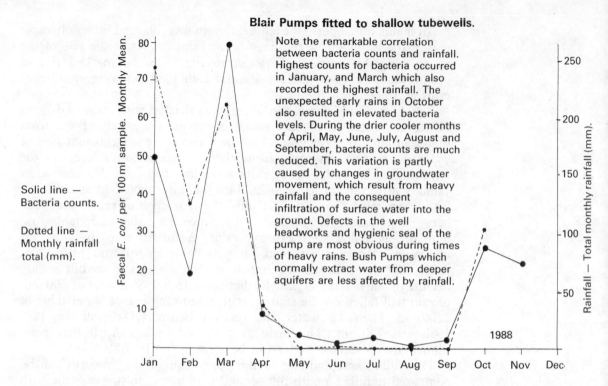

Figure 3. Seasonal variation in bacteria counts.

This situation is revealed by studying the quality of groundwater raised by handpumps through the seasons. Fig. 3 shows the seasonal variation of water quality from a number of Blair Pumps in Epworth. The highest counts for bacteria occur during the rainiest months, January and March, the lowest counts during the dry season. A similar rise and fall in the quality of water taken from Upgraded Wells in Epworth is shown in Fig. 2. This figure also clearly shows how the heavy rainfalls of January and March are followed by rapid rises in the water table, and corresponding increases in the numbers of bacteria found in water samples. During February, the rainfall was less, the groundwater table far more static and bacteria counts in water samples taken from both pumps and wells much reduced.

These data show very clearly that the number of faecal *E. coli* taken from shallow groundwater sources is elevated after periods of rain, when the groundwater is moving at its fastest through the soil. It is clear that the contaminants in the ground are moving at their fastest when the vehicle carrying them, i.e. water, is also moving at its fastest. These movements will occur during and after periods of heavy rain, when the water table is moving horizontally and also gaining height at its fastest rate.

However these are not the only influences of rain on groundwater quality. Poorly made concrete headworks (apron and water run-off) can crack, and will allow leakage of waste water from the surface back into the well or borehole. Similarly where handpumps are loosely fitted and worn in such a way that water can drain from the apron through the pump head into the well, then contamination of the well water is inevitable. These weaknesses in design or construction are revealed most dramatically during rainy spells, when the rain helps to flush surface contaminants back into the well or other water source. The hygienic seal of the water point is a most important feature of a well head and is tested most thoroughly during the rains. It is at this time when contaminants from the surface most commonly find a route and drain back into the underground aquifer.

Hand-dug wells

Traditionally, man has always extracted water from the ground through hand-dug wells. This method is supremely successful and has been used throughout the world for thousands of years. It is estimated that in Zimbabwe alone nearly 100,000 wells or water holes may be in operation daily for most of the year, and it is certain that more water is gathered from this source compared with any other in the rural areas of Zimbabwe.

Shallow wells are used by more people than any other single source of water in rural Zimbabwe.

Shallow wells

In its simplest form, the well is little more than a water hole, which is hand-dug down into the water table. In the simplest shallow wells, steps are cut into the side walls to facilitate entry from above. Alternatively a bucket and rope are used to raise water to the surface. Such wells are very common in many rural areas, but are subject to gross contamination and very often yield water of unacceptable quality. During the rainy season, run-off water from the surrounding areas drains into the water hole or well, carrying with it many forms of contamination. The water-logged conditions surrounding many traditional wells and waterholes are subject to contamination by the passage of bacteria from surrounding areas on the feet of users and on animals which may be attracted to the surrounding water pools. This pollution is further carried into the well on buckets and ropes which are used to raise the water, and often lie around the unhygienic rim of the well. Furthermore animals can fall into open wells of this type, which are also dangerous for children.

Traditional methods of upgrading shallow wells

Many steps have been taken in traditional practice to improve the safety and permanence of wells and the quality of water they yield. Well collapse can be avoided by lining the walls of the well with stones or burnt bricks. In addition, coverslabs made of wood or concrete prevent the danger to children and reduce the chances of animals and foreign objects falling into the well. This technique is enhanced by the addition of a well lid or cover.

Of equal importance is the technique of raising the well head above the surrounding ground level to divert run-off water away from the well site. The addition of a hygienic apron and water run-off is of the greatest significance, as this diverts waste water away from the well head into a nearby seepage area.

One of the most significant steps used to improve water quality is the windlass in combination with a chain and bucket. Not only does this simple technique make bucket lifting easier, but it also allows for the chain or rope to be hygienically wrapped at the head of the well. Without a windlass, the rope or chain becomes contaminated in the wet surroundings of the well head. Contamination of the bucket itself is reduced if it can stand on a raised concrete or brick collar surrounding the well, when not in use. Buckets not stored in this way pick up contamination from the unhygienic well head, and repollute the well once they are lowered to collect water.

Contamination of the bucket is also reduced by hanging it on the windlass handle or a special bucket hook attached to the windlass support. When the bucket hangs free, the outer surface dries off quickly (unless it is raining) and often gets hot if left in the sun. These effects dramatically reduce the number of bacteria carried on the bucket.

The windlass. One of the most ancient and successful ways of raising water.

Deep wells

The shallow wells so far described, are dug into soils with a pick and shovel and generally are not more than 15 metres deep. They do not penetrate the harder rocky layers found more deeply in the ground.

Deep wells have also been excavated in Zimbabwe for many years, mainly in Matabeleland, where they can attain depths of over 30 metres. Within the last few years the number of deep wells has increased considerably and over 4000 may now be in existence. Deep wells require greater expertise to dig, since they penetrate much deeper aquifers which often lie in rock formations. Often penetration of the ground is achieved with the use of explosives, and is a specialised task performed by skilled well sinkers. Well lining techniques are more sophisticated than those used in shallow wells, since the impact of the explosive must be borne. Being more expensive to construct (usually about Z$2500 (US$1250) in 1988) they generally serve larger communities, rather than families, and are fitted with robust handpumps like the Bush Pump. Large numbers have been built with assistance from the Lutheran World Federation and UNICEF.

Excavating a deep well.

Building an upgraded shallow well

Whilst many wells used in the rural areas may have been built with some protective features, very few are built with all the features combined and thus still yield water of rather poor quality.

It is easy to build an upgraded well, either by constructing in a completely new site or by improving on an existing facility, which may have been used for many years. The advantage of choosing an existing site, is that much of the hard excavation work will have already been completed. However some existing wells may have been sited poorly, and this must be taken into consideration before the well is upgraded. Many traditional wells were dug deep in times of drought, and thus have become very reliable sources of water. If they were also lined with good quality fired bricks, they offer an excellent starting point for finishing off the upgrading process.

Upgraded wells can be built using traditional skills and materials, cement being the main imported ingredient. They are usually cheap to construct and easy to maintain. If they are used carefully they can yield water of good quality, without the need of a handpump, which greatly reduces the costs of maintenance. They are best dug during the dry season, when the water table is at its lowest, with the months of October and November being the best in Zimbabwe. They have a very long life and in suitable ground can be built close to where people live. Under such conditions, more water is used for cleaning, bathing and for gardens, all of which have beneficial effects on family health. If further improvements in water quality are desired, the upgraded well can be protected further by the addition of a suitable handpump.

Well siting

The final quality of the water taken from a well will be partly dependent on how close it is to potential sources of contamination.

The well site should be placed on raised ground and at least 30 metres away from latrines, cattle kraals and refuse pits or other hollows in the ground. This ensures that contaminated water will not drain through the ground into the well. It is important that the well is always uphill of a latrine or cattle kraal. Obviously the site should be close to where people live and be convenient for them. The precise site should be chosen by the community in consultation with a Health Assistant, and with the help of a local water diviner, who may be able to locate a site with a high yielding aquifer. The diviner will look at the lie of the land, make a note of the soil type and catchment area, and look for other signs such as '*Muonde*' or '*Mukute*' trees, or lines of anthills. His general knowledge of the land coupled with the movement of the divining stick may locate a good site for water. If a band of capillaries or underground streams is found, the well will have a good yield compared with one that penetrates clay.

Excavations

1. Existing wells

If an existing well has been chosen for upgrading this should be deepened as far as possible and the sides cut straight. Clearly this is best achieved during the time of the year when the water table is at its lowest (October and November). An experienced well digger and his team should be chosen to excavate the final few metres of an existing well. A windlass should be used to raise and lower the men and buckets into the well. If possible two smaller (20 litre) or one larger (50 litre) bucket should be used for dewatering and final excavation. The reliability of the well will depend on how deeply it penetrates into groundwater. A supreme effort must be made at the final stage of digging. The inflow of water increases as the well is deepened. Once the inflow is high, and the water can no longer be extracted with sufficient speed to keep the bottom exposed for digging, the excavation must stop.

2. New wells

After a suitable site has been chosen, a 1.5 metre diameter ring is marked on the ground to show where the well should be dug. Less experienced labourers can dig the first few metres of the well which should be excavated with straight sides. As the well becomes deeper more experienced well diggers should continue the job. The excavation is best carried out with a stout windlass and bucket together with a team of at least three men. Steps can be cut in the walls of the well to assist in climbing out. The diameter of the excavation can be reduced to 1.2 metres in harder rocky layers which do not require lining.

Once the water table is reached, the excavation continues, with more water being extracted with the soil (cuttings). As the digger continues, the number of buckets of water drawn out becomes greater in comparison with the number of buckets of harder material. At some stage the rate of inflow of water into the excavation equals the amount that can be drawn out by the bucket and windlass. Two buckets may be used at this stage — but there comes a time when the water inflow is too high and further digging is impossible. Well digging may also become difficult in loose collapsing soils, in which case it may be necessary to dig within a series of concrete well liners in order to penetrate the aquifer (see later). On the final day, a supreme effort must be made to penetrate as far as possible into the water. Normally it is possible to penetrate two or three metres into the aquifer. If there is a chance of collapse of the well, the well lining should be built as soon as possible.

Mark a 1.5 metre diameter ring on the ground.

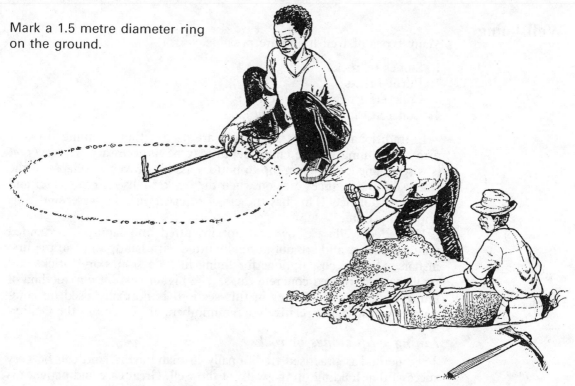

The well is dug down with straight sides to the water level and then as far below the water level as possible.

Continue to dig, even when water is first reached.

Well lining

Many types of well lining are possible:

1. Stones or rocks
2. Burnt bricks
3. Concrete rings (prefabricated)
4. Concrete rings (cast '*in situ*')

In traditional practice, many wells are used without a lining. In very firm soils, lining for the entire depth may be unnecessary. However, it is wise to line a well from top to bottom as this prevents collapse of the side walls and reduces erosion when the bucket falls into the water and causes turbulence. This has the effect of clarifying the water and thus improving taste.

Upgraded wells are low cost options fitted into family or extended family settings, and are not normally fitted with handpumps in the first instance. Lower cost options for lining include stones and bricks and smaller prefabricated concrete rings. The higher cost '*in situ*' method of lining wells described later in this section, is normally used on community wells, where relatively large numbers of people use the facility.

Lining with stones or rocks

This method is practised traditionally in many areas, and can be very successful at holding up the walls of the well. Great care and patience is required to stack the rocks without mortar so they form a rigid tube. Where rocks are stacked above water level they can be held together with cement mortar. When using rocks as a well lining it is wise to pack puddled clay behind the rocks to form a seal so as to reduce erosion of the well behind the rock layer, and so reduce seepage of pollutants from the surface. A good hygienic seal is essential at the head of the well. The well lining must be cement mortared nearer the surface and should be built up 300 mm above ground level. Puddled clay should be well packed in the annular space between the excavation and the rock lining. The uppermost section should be well mortared and levelled flat in preparation for the well cover.

Lining with burnt bricks

This is one of the most common methods of lining wells in the rural areas. The bricks should be well fired, otherwise they will crumble when under water.

Below the water level, the bricks are carefully stacked without mortar to form a tube at the base of the well. This section should be supported by a packing of gravel between the brickwork and the wall of the well. Cement mortar should be used with the brickwork as soon as it is technically possible and certainly above the level of the water. Where there is some doubt about the stability of unmortared bricks, a lower lining of flat rocks can be stacked at the base of the well and mortared brickwork built upon this lining. The mortared brickwork is built up to 300 mm above ground level. If possible, puddled clay is packed between the excavation and the brickwork to form a sanitary seal, especially nearer the head of the well. This helps to prevent contaminated water trickling into the well from the surface.

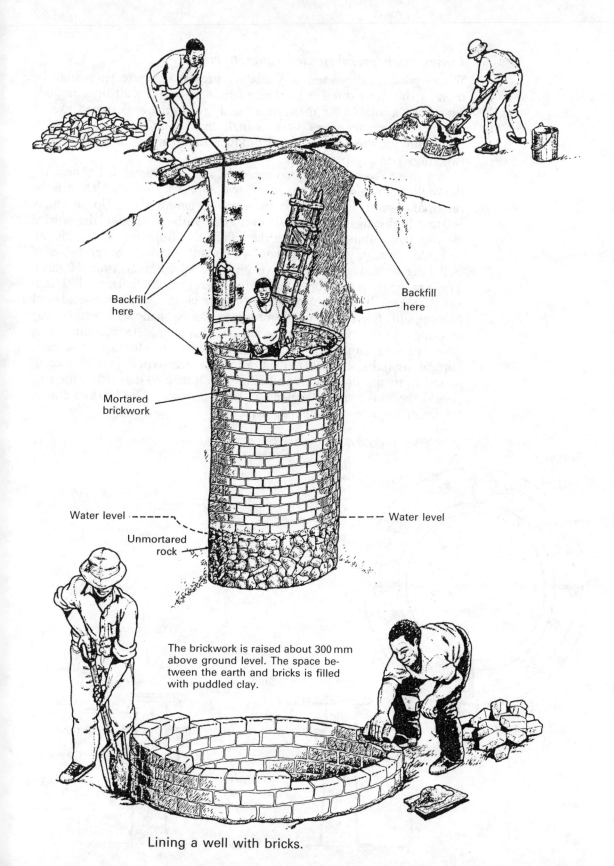

Lining a well with bricks.

Lining with prefabricated concrete rings

Where good quality sand is available, precast concrete rings may be used as the most durable method of lining a well. Well liner moulds have been available for many years and are normally designed to cast 1.2 metre diameter rings 300 mm high and about 75 mm thick. With this technique the normal rate of production of rings is one, or at most, two rings per day per mould.

In a more recent technique developed at Blair Research Laboratory, the well liner mould is made of two tapered shells which can be removed almost immediately from the concrete ring. Up to thirty 300 mm high rings can be made every day with the mould, the normal number being about twenty per day. Thus the lining for most wells can be made, on site, within two days with a single mould. Several sizes of well liner mould are available (from V & W Enginneering, Harare). These range from 600 mm internal diameter, through 800 mm, 1000 mm and 1200 mm. Using a mixture of 5 parts clean sharp river sand to 1 part cement, five complete 600 mm rings can be made with a single bag of cement. One bag of cement makes four complete 800 mm rings. Three or four loops of 3 mm wire are used as reinforcing. The most popular mould is the 800 mm internal diameter type, since it is economical in its use of cement, and allows for a man to dig within the ring should the well require deepening at a later stage. The 600 mm ring is too small for a man to operate a pick.

Well liner mould.

Technique

The well to be lined is deepened as far as possible during the months of October and November, when the water table is at its lowest. It should contain at least two metres of water at this time of the year.

The depth should be measured and the exact number of rings calculated, one per 300 mm (1 ft) depth, plus one ring above ground level. For a 6 metre well twenty two rings are required.

To prepare the rings, an area of ground is levelled flat and large enough to accommodate all the required rings. Sufficient bags of cement are collected to make the rings, together with a suitable quantity of river sand. If 800 mm rings are to be made, one bag of cement is required for four rings.

Clean the mould and paint old engine oil on the inner surfaces of the two shells. Assemble the mould on flat ground. The concrete is prepared by mixing one bag of cement with five times its volume of river sand (1:5 mixture) and sufficient water to make a dry (stiff) mixture of concrete. This mixture should hold the least possible amount of water to make good concrete, and some experimentation should be carried out. If the mixture is too wet, the ring will slump after the mould is removed.

Add the mixture between the shells of the mould to a depth of 75 mm and then firmly ram down with rods or wooden sticks. Add a ring of reinforcing wire. Add a further layer of concrete mixture and ram down again. Add a further ring of reinforcing wire. Fill the mould completely with the mixture ensuring that it is packed with a solid mixture of concrete. Level off the top of the mixture.

The mould can then be removed. Take out the spacers and remove the inner shell first, very carefully, followed by the outer shell. The shells must remain completely level when being extracted. The concrete ring should stand up by itself without collapsing. The mould can be cleaned and oiled, and the next ring made in the same way. One mixture of concrete using one bag of cement is processed into rings at one time. The rings are covered with hessian or paper to protect them against the sun or rain. They are kept wet for one week in sheltered conditions.

Lining the well

When using this technique, the well liners are lowered down the well one after the other with the broad base of the liner downwards. During the moulding process three or four holes are left in the wall of the ring, through which wire can be looped and rope attached for lowering.

Care should be taken to ensure that the lowest ring is placed centrally and level on the well bottom. A further two rings are added on top of the first. A sheet of plastic is then carefully wrapped around the rings to cover the joint. The annular space between the rings and the well is filled with either thoroughly washed river sand or cuttings from the well excavation. Since the rings do not have a stepped joint, the plastic stops sand running through the joints between the rings.

A further three rings are stacked and the same procedure followed.

Once above the water level, the rings can be cemented together. The annular space between the well and the rings should be filled with cuttings from the well, which should be uncontaminated. The rings are further stacked until at least one ring stands above ground level. The uppermost half metre of annular space should be filled with concrete. This will form a sanitary seal, a foundation for the apron and an anchor for the windlass supports.

When the finished well is to be fitted with a handpump, the concrete backfill can be built up to ground level, and an apron built over this. However when the well is to be fitted with a windlass, two wooden poles must be anchored in the concrete, one on each side of the well lining, and provision must be made for these in the concrete backfill surrounding the rings. Normally, poles are embedded about 700 mm in a concrete anchor. Part of this will be anchored in the concrete apron itself (200 mm), and part in the concrete backfill (500 mm).

It is possible to make rings with special stepped joints which fit one on top of another. These joints make a better seal, especially below water level where joints cannot be cemented together. In this case the mould is equipped with two rings which are set in such a way that they leave a step in the upper and lower faces of the ring. Jointed rings can be made with the tapered mould by inserting suitably cut rings of polypipe so as to make the recesses on the upper and lower surfaces of the ring.

Tapered moulds and step jointed moulds are available from V & W Engineering, Harare.

Well lined with precast concrete rings.

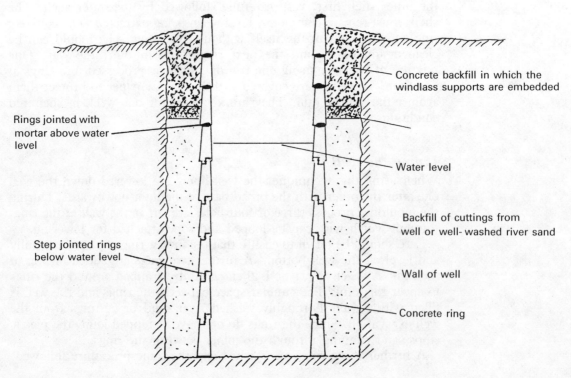

Lining the well with concrete rings cast 'in situ'

This technique was developed by the Lutheran World Federation in Matabeleland, and is now commonly used in several parts of Zimbabwe. The technique is fully described in a booklet available from DDF (Water Division, Harare) and UNICEF (Harare).

In this technique, which can be used to line wells down to more than 30 metres, the well sinker excavates a well of 1.5 metres diameter down to the layer of harder rock, where blasting is required. After this depth the well is excavated 1.2 metres in diameter. Before blasting commences, the upper 1.5 metre section is lined with concrete which is cast '*in situ*' using steel shutters. The shutters, which are 1 metre deep, are first assembled on the shoulder of the lower 1.2 metre section of the well. A concrete mixture consisting of cement, stone and river sand is mixed and added to the annular space between the well and the shutter. Two bags of cement are normally required per metre section of lining. The mixture is left to set overnight, and in the morning the shuttering is removed, cleaned, oiled, and relocated on top of the previous lining and a further one metre of well is lined. Some experimentation is required with locally available materials to arrive at an economical mix which retains strength. The same technique continues to the surface. An outer shutter is used above ground level so that a ring approximately 500 mm high is formed at the well head.

Once the upper well lining has set, the well can be blasted through the rock layer and through the water-bearing aquifer to the required depth. In community wells fitted with a handpump, at least one hundred 50 litre mining buckets of water are required per day, before the well is considered deep enough for community use.

A windlass is essential to excavate medium depth and deep wells.

The mining bucket used to excavate wells.

The *'in situ'* method of lining wells is not normally used for upgrading existing wells, and may rarely be used as a technique for family wells because it is relatively costly, and heavy specialised equipment is required. However, it is an excellent method of lining wells where funding permits. The final choice of well lining can only be made when local conditions are known.

'In situ' well liner mould.

The diameter of the well is reduced from 1.5 metres to 1.2 metres when harder rock is met. The upper softer material is lined with concrete, the lower harder material is unlined.

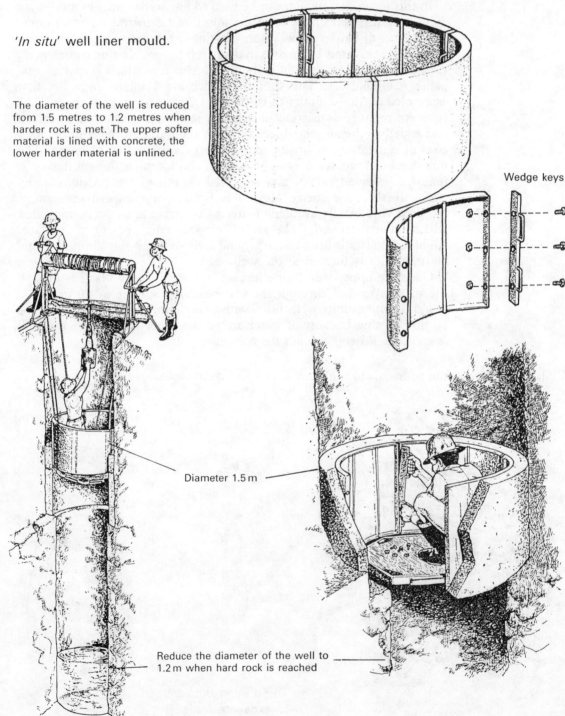

Wedge keys

Diameter 1.5 m

Reduce the diameter of the well to 1.2 m when hard rock is reached

The well cover

Once the well has been lined, the next stage of protection involves fitting a concrete well cover. Many poorly protected wells built in the rural areas have little more than a few logs placed across the opening of the well as a precautionary measure to prevent accidents with children. In more advanced traditional techniques, logs are placed across the head of the well and these are mortared in position with concrete. Where this technique has been carried out above ground level, the well can be regarded as semi-protected, at least against run-off water.

Making the well cover

All upgraded or protected wells are fitted with a concrete well cover which is cemented to the uppermost well lining. The cover slab is important because it helps to prevent polluted waste water and other objects from falling back into the well. It also makes the well safer for children and provides an hygienic resting place for the family bucket.

The diameter of the slab will depend on the well size and should fit neatly over the upper well lining. A hole is left in the slab through which a bucket can pass or a pump can be fitted. The exact size of the hole will vary depending on the type of facility to be fitted to the well head. In the case of the upgraded well, described here, the hole is made large enough for a standard 10 litre bucket to pass through it. This will normally be about 325 mm in diameter.

Technique

An area of land close to the well is cleared and flattened. The external diameter of the well collar is measured and a circle of the same diameter is marked on the ground. This will depend on the well lining technique — for an 800 mm internal diameter well liner, the slab diameter will be 900 mm. The size of the internal hole will be about 325 mm in diameter to allow a 10 litre bucket to pass through easily.

The ring of bricks which forms the mould is laid around the marked circle and a suitably sized tin placed centrally within the ring. Alternatively and preferably a special steel mould about 325 mm in diameter and 75 mm deep should be used to make the central hole. Some 3 mm reinforcing wire (8 gauge) is now cut and placed within the mould to form a grid with 150 mm spaces.

Mark out the cover slab to suit the well.

A tin drum can be used to make the central hole in the slab. A 10 litre bucket should easily pass through this.

The wire is removed, and the concrete mixture made up. If small stones are available the mixture should be 3 parts stone, 2 parts river sand and 1 part cement. If stones are not available the mixture should be 4 parts river sand and 1 part cement. The concrete is added to half fill the mould. The reinforcing wires are then added and laid in a grid formation 150 mm apart. The remaining concrete is then added and rammed down hard and wood floated flat. The final thickness of the slab should be 75 mm.

Adding the protective collar

The final opening of the upgraded well will be fitted with a tin cover for maximum protection. This fits over a collar which is built on the cover slab and is made of concrete. Bricks can also be used if preferred. In order to get the size of the collar correct it is wise to have the tin cover made first and then shape the collar to suit the cover. Tin covers are normally made about 500 mm in diameter and 75 mm deep, and are fitted with handles.

The concrete cover slab is left to cure for about one hour, after which the central mould can be removed. The central hole is then filled with wet sand.

The central mould is then replaced on top of the sand, and a ring of bricks laid around the mould so that a collar 75 mm high and 75 mm thick can be made around the central hole. The space between the bricks and the mould should be filled with more concrete and levelled flat and left to cure for a further hour.

The ring of bricks and the steel mould (or metal tin) should now be removed and the concrete collar shaped and smoothed with a steel float so that the tin lid will fit neatly over it.

The completed well cover should now be left to cure for at least three days and kept wet at all times. If a stronger concrete mixture is used and the well cover is not too large, it is possible to move the cover, with care, after 24 hours.

The well cover should be carefully lifted, and laid in a bed of cement mortar over the raised brick well lining (or concrete rings if these are used). The exposed brickwork should be neatly covered with a layer of cement mortar.

A central collar is built up in concrete around the central well cover hole. The inner mould can be made from a can and the outer mould from bricks.

When it has begun to cure, the central collar is shaped so that the tin lid will fit neatly over it.

Bricks can also be used to make a central collar. This is best made after the cover slab has been fitted to the well.

The Upgraded well. A raised lip on the cover helps to stop water draining back into the well.

35

Adding the windlass supports

In the upgraded well, water is raised with a bucket and chain fitted to a windlass. If a pump is fitted to the well at a later date, the windlass and its supports can be removed.

Normally windlass supports are made of stout wooden poles 125 mm to 150 mm in diameter and 1.8 to 2.0 metres long. Ideally these should be treated against termite attack, if they are not already resistant. The poles are embedded in a concrete anchor on either side of the well liner so that they project above the final apron level by about 1.2 metres. If the well is lined with small concrete rings the final concrete backfill behind the rings is added once the poles are placed in position. The total depth of the concrete anchor should be about 600 to 800 mm deep including the apron. This is illustrated in the diagram. Because wood has a tendency to expand when wet, some lengths of 3 mm reinforcing wire should be arranged around the poles in the concrete to strengthen the concrete at these points, as it is liable to crack. This is an important part of the assembly, and cracks should be avoided, since waste water could leak back from the apron into the well, leading to contamination.

Once they are firm, slots can be cut in the upper end of each pole to accept the steel windlass. These should be about 100 mm deep, and be cut a little wider than the windlass shaft itself.

Two windlass supports are embedded in concrete one on each side of the well.

The windlass

The windlass is made to fit the well. Where an existing well has been upgraded, a windlass may have existed before and this can be used again if it is well-made. However, a new well-made windlass is preferable and these are available commercially. A windlass is provided as part of a subsidy in recent well upgrading experiments. Obviously a new windlass will be required for a new well.

Normally windlasses are made with a long steel shaft fitted with a handle at one end. The barrel of the windlass is a section of thicker pipe, which is normally welded to the middle section of the shaft. In rural conditions, where welding equipment may not be available, a piece of 20 mm or 25 mm steel water pipe or steel rod can be used as a

Thousands of traditional wells in Zimbabwe are semi-protected with a well cover and windlass.

shaft and bent at the end to form a handle. If a larger pipe (50 mm or 75 mm) is available this can be attached to the shaft by filling the space between the shaft and the pipe with a strong mixture of concrete. Alternatively a wooden barrel can be fitted over the steel shaft and held in place with steel straps or wire. There are many traditional methods of making a windlass. The exact technique is not critical, what is important is that it makes raising the bucket easier, and provides a hygienic surface around which the chain or rope can be held and stored.

The windlass is made to fit the well.

The apron and water run-off channel

This is an important part of the upgraded well and should be made with care and in strong reinforced concrete. It should be at least 2 metres in diameter and preferably 3 metres. The water run-off should be made at least 4 metres long and preferably 6 metres long for the family well. The main purpose of the apron and water run-off is to carry waste water away from the site of the well into a seepage area, which should be sited slightly downhill of the well, to facilitate good drainage.

The concrete is laid to a depth of 75 mm to 100 mm within a circle of bricks or rocks which forms the rim of the apron. The outer rim should be raised further over the rocks or bricks to a depth of about 150 mm. The upper surface should be shaped so that it is smooth and allows all the waste water to drain into the run-off channel. The entire apron should be reinforced with 3 mm wire to prevent cracking which would ultimately lead to repollution of the well. As mentioned earlier, it is also advisable to place a few steel reinforcing wires around the wooden windlass supports where they pass through the apron. This helps to prevent the concrete cracking at this point. An additional layer of cement mortar should be added to the joint between the lining and the apron as shown in the diagram to avoid the accumulation of waste water at this point.

The water run-off channel is made in the same way, with bricks being laid down as a base to form a channel through which the water can pass. It is important to slope both the apron and run-off channel so that water can completely drain away. Both apron and channel should be strongly made to avoid future cracking. Many aprons and channels are poorly made and crack easily after a year to two's use. It is important that the apron and water run-off be left to cure slowly and should be kept wet for several days after construction.

The seepage area

This can be a hollow in the ground filled with rocks or gravel or plants like bananas. In the case of the upgraded well this will usually be built in a homestead, in a central place. Under these conditions, the ideal setting for the seepage area is within the vegetable garden. It is possible

Upgraded wells make use of traditional concepts. The windlass, tin lid, and apron all exist in traditional practice. They are combined here to make a safe unit for the household.

The Upgraded well. Indigenous hardwoods last a long time as windlass supports. Many materials can be used in place of a chain to raise the bucket.

to build a sump at the end of the channel, into which waste water can flow, and from which the garden can be watered. Even waste water is valuable and should be put to use.

The bucket and chain

A durable 10 litre bucket and chain should be fitted to the windlass. Ideally the bucket should be weighted on one side so that it tips up and fills with water easily. A builder's bucket is more durable that the standard tin type. Chains are more hygienic than rope, and should be used if possible and fitted to the bucket and windlass with wire. Alternatively 3 mm or 4 mm wire cable can be used, as this may may be cheaper than chain.

Correct use of the upgraded well

The upgraded well can provide water which is far purer than ordinary well water, provided it is used with care. When the installation is finished, health workers should advise the users how they can best utilise the facility. The following points are most important.

1. Keep the bucket clean.
2. Hang the bucket on the windlass when not in use.
3. Keep the well lid in place.

The finished 'Upgraded well'.

4. Keep the apron, run-off and seepage area clean.
5. Always use the same bucket in the well.
6. Keep the chain wrapped around the windlass.
7. Store the water in clean covered containers.

An illustration of a leaflet (in English) given to users of upgraded wells is shown in the diagram on the next page.

Cross-section of completed brick lined 'Upgraded well'.

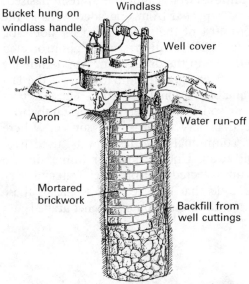

Near the bottom of the well it is difficult to cement mortar bricks together, although this can be attempted. Unmortared bricks should be carefully stacked and mortared as soon as it is possible above the bottom. Carefully stacked rocks can also be laid at the base of the well. The backfill should be made from well cuttings near the bottom.

Cross-section of completed concreted ring lined 'Upgraded well'.

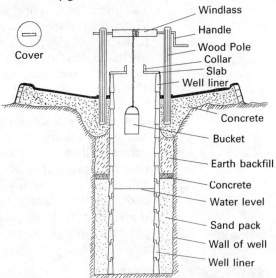

Note: There is some variation in the backfill. This can successfully be made with the cuttings from the well itself or with well-washed river sand. Soil from the surface should not be used for the lower backfill. Soil can be used nearer the surface.

How to look after your well

You can drink clean water from your Upgraded well by keeping the well and bucket clean

1. Keep the bucket clean
2. Hang the bucket on the windlass
3. Keep the well cover in place
4. Keep the apron and run-off clean
5. Always use the same bucket in the well
6. Keep the chain wrapped around the windlass

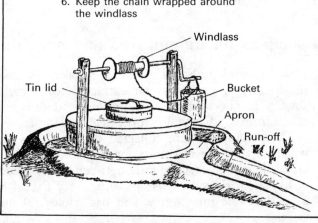

The importance of the upgraded well

The importance of upgrading traditional wells cannot be overemphasised. Water of improved quality can be expected from upgraded wells as the tables below clearly show. About 40% of the water taken for domestic use in the rural areas of Zimbabwe (1988) is withdrawn from unprotected wells in the wet season and about 30% in the dry season. Approximately 100,000 wells or water holes have been dug in Zimbabwe over the last half century and represent a valuable national resource. These figures show that wells are an important source of water in rural Zimbabwe, and are used by more people than any other single source. Where family wells are very common, communal facilities may be used less, simply because they are further away. Under these conditions disease will result if the family well is unprotected and provides water of very poor quality. Clearly any technique that can be used to make these wells safer to drink and safer for children should be encouraged and put into widespread practice.

Water in upgraded wells should be clear and fit for drinking and taste good.

Wells offer an important source of water in rural areas.

Bacteriological samples for unimproved and upgraded wells

Blair Research Laboratory has undertaken extensive surveys in which the quality of ground water extracted for domestic use from various sources has been tested. The following data is extracted from records taken from studies at Epworth near to Harare.

These samples were collected during the wettest time of the year when bacteriological counts for all protected and unprotected sources are at their highest. The ratio between unprotected and upgraded wells is 4.07:1 for Faecal *E. coli* and 5.62:1 for Faecal *Streptococci*, reflecting a considerable improvement in bacteriological quality. Improvements

Table 1. Faecal *E. coli* per 100 ml sample collected from eight unprotected and eight upgraded wells 4.1.88 – 23.3.88.

	UNPROTECTED WELL	UPGRADED WELL
MEAN	342.48	84.01
NO. SAMPLES	85	86

Table 2. Faecal *Streptococci* per 100 ml sample collected from eight unprotected and eight upgraded wells 11.1.88 – 23.3.88.

	UNPROTECTED WELL	UPGRADED WELL
MEAN	579.48	103.01
NO. SAMPLES	88	88

in turbidity were also noticed after the upgrading process with a related improvement in the taste of the water.

During the winter months (May to August) bacteriological counts fall as the weather becomes much drier and cooler. At such times, the movement of potential contaminants both on the surface and in the ground is much reduced due to the scarcity of rain. This is because groundwater moves less quickly through the overburden, and is less able to carry bacteria for any distance to the site of the well. Similarly on the surface, reduced rainfall limits the flow of run-off water from surrounding areas which may also contaminate the well. The long periods of direct sunlight, resulting from reduced cloud cover, have a sterilising effect on the bucket, chain and other exposed components of the well head. The drier conditions on the apron and its surroundings also allow for improved hygienic conditions at the well head, and these greatly influence the quality of water in the well itself.

These data show the considerable advantage in upgrading a well for family use. Once upgraded, simple wells of this type can provide long periods of service with the minimum of maintenance. Their similarity to wells used in traditional practice ensures that they are understood and can be maintained with ease at village level, and particularly in the homestead, where they are best placed. Indeed this system lends itself perfectly for installation at the family level, where the highest degree of care and maintenance can be expected.

Whilst it is true that bacteria counts are higher than in wells fitted with handpumps (see below), a handpump supply is only safe whilst the pump is operational. Should the pump break down without immediate repair, villagers are often forced to take water from totally

Table 3. Faecal *E. coli* and Faecal *Streptococci* per 100 ml sample collected from Blair Pumps and Bush Pumps fitted to wells and tubewells in Epworth 6.1.88 – 28.3.88 (during the rainy season). SOURCE: Blair Research Laboratory.

	Feacal *E. coli*	Faecal *Streptococci*
MEAN/100 ml	35.92	15.07
NO. SAMPLES	177	169

unprotected sources, often with disastrous consequences. In practice the upgraded well offers a much more reliable source of water, especially for a family where the number of users is small, and the degree of care is high. Under such conditions, the users can more easily harmonise with the well itself, including the micro-organisms that may pass too and from the well via its users. Where the number of users is high, as in most community water points, the potential for complex cross-contamination of an open well is far higher and the same principles can no longer apply. A suitable handpump, which ensures greater protection of a well from foreign sources of contamination, should always be fitted in a community setting, provided that it is established with a working maintenance system.

Handpumps offer a greater degree of protection because they should seal off the well from external sources of contamination. However even when handpumps are fitted, contamination can still pollute the well through the headworks if the apron or water run-off is cracked or the well is badly sited. Fitting a pump to a well does not guarantee, in itself, that high quality water will be produced.

The protection of wells and tubewells with handpumps is described in greater detail in the following chapters.

Table 4. Distribution of drinking and domestic water sources by type and source. SOURCE: Piers Cross, Zimbabwe National Master Water Plan, 1988.

IMPROVED SOURCE	WET SEASON %	DRY SEASON %
Household tap	0.9	0.8
Community tap	4.6	4.8
Protected well	7.4	6.3
Borehole	19.0	26.2
Protected spring	0.7	0.6
UNIMPROVED SOURCE		
Unprotected well	41.4	31.4
Unprotected spring	7.0	7.5
River	10.5	9.2
Dam	1.6	3.5
Rainwater collection	0.2	0.0
Sand abstraction	5.5	8.3
Other	1.2	1.4
	100.0	100.0

NOTE: These figures show that in the wet season 67.4% of the rural population take their water from unprotected sources (61.3% in dry season). Only 32.6% use improved sources in the wet season and 38.7% in the dry season.

Implementing 'upgraded well' programmes

In Zimbabwe it is now becoming increasingly apparent that in areas where family wells are very common, the advantage of fully protected communal water points may not be fully realised. Very often, although a well or borehole fitted with a handpump is available in the area, people still prefer to use their own family well simply because it is close at hand.

Most family wells are poorly protected, and yield water of poor quality for most of the year, and this can lead to outbreaks of enteric disease, even in areas where improved water sources are available but at some distance from the household. Clearly situations like this cannot be ignored and must be addressed, the most practical method being to implement family well upgrading programmes.

Family wells exist in very large numbers in many parts of Zimbabwe, the total number exceeding 50,000. Whilst families are being encouraged to upgrade their own wells completely by themselves, recent pilot schemes in the Makoni District of Manicaland and the Goromonzi District of Mashonaland East have used a slightly different technique in which each family is given a subsidy as an incentive to improve its own well.

In these cases the family subsidy has been chosen to include a mass-produced windlass, a tin lid for the well cover and sufficient cement to perform the concrete work, usually three or four bags. The value of this subsidy amounts to Z$50–Z$60 (U$25–US$30). The family is expected to provide bricks, labour, sand and stone, a bucket and rope and the windlass support poles. If the family wishes, it can further protect the well by providing a handpump, like the Blair Pump which is suitable for family use.

The pilot schemes are being introduced on an experimental basis to prove the viability of the concept as a means of widespread implementation, and also to demonstrate the health impact of family based water programmes. This move has been taken following the success of the Blair Latrine Programme which is also based on a family subsidy. Family well upgrading schemes will be linked with the promotion of family gardens and health education programmes which emphasise the importance of personal hygiene in the homestead.

Where wells are unlined, the family is encouraged to deepen them and line them with bricks or cement rings. Where wells have already been lined with bricks, deepening and further lining are also encouraged, but the emphasis is placed on improving the well headworks. This includes raising the well lining above ground level, fitting a concrete well cover with raised collar for the tin lid and fitting the windlass to its supports. In addition a strong apron and water run-off channel are built. The importance of adequate curing by keeping the concrete work wet for several days is also emphasised. The family is encouraged to buy a new bucket if the old one is worn out and instructed in the correct hygienic use of the well. The family takes complete responsibility for well maintenance: an operation which must be undertaken by Government on most communal water supplies.

Hand drilled tubewells

Although the wide diameter well is the most common type of excavation for penetrating shallow water tables in Zimbabwe, and most other developing countries, this is being replaced more and more with narrow diameter wells called tubewells. Most boreholes drilled with mechanised rigs can be described as tubewells, and these often penetrate into very deep aquifers where water is found in fractured rock layers.

Where water is abundant in the 'overburden' above the rock layer, it is possible to drill a shallow borehole or tubewell by hand, provided the ground is suitable. If the water table lies within about 6 metres of the surface, and the ground is fairly soft, this can be drilled with a standard post auger attached to standard 25 mm water piping and turned with wrench spanners. Several hand-operated drilling rigs have been designed throughout the world, but one of the best was designed by Mr E. Von Elling of V & W Engineering in Harare. This machine, called the Vonder Rig, represents a significant advance in hand-operated drilling rig design, and has pushed the state of the art very much forward.

The Vonder Rig is illustrated on the next page and consists of a sturdy tripod from which the drilling stems are slung. The heart of the rig is a robust worktable, which aligns the drilling auger, and makes possible the drilling of vertical tubewells down to a depth of over 35 metres in the ground. The auger itself consists of a steel tube fitted with hard steel blades, for cutting and lifting soil. The auger is attached to the lowest drilling stem by means of a bayonet adaptor. The auger and drilling stems are raised and lowered with a winch which is attached to one leg of the tripod. Several types of auger are available, one being a hole saw which can penetrate decomposing rocks such as decomposing granite. The Vonder Rig was designed to drill through soils and decomposing rock formations, but not through hard rock. The manufacturers,

V & W Engineering, P.O. Box 131, Harare, Zimbabwe, supply the following items with each drilling package:
1. Worktable with plumb line for levelling
2. Tripod with hand winch and cable
3. Auger (for soft soils)
4. Hole saw (for decomposing rock formations)
5. Bayonet adaptor for augers
6. Cross bar for turning stems
7. Bailer
8. 8 × 2 metre heavy duty drilling stems with acme threads
9. Stem stand with oil can
10. Heavy duty spanner for tightening stems

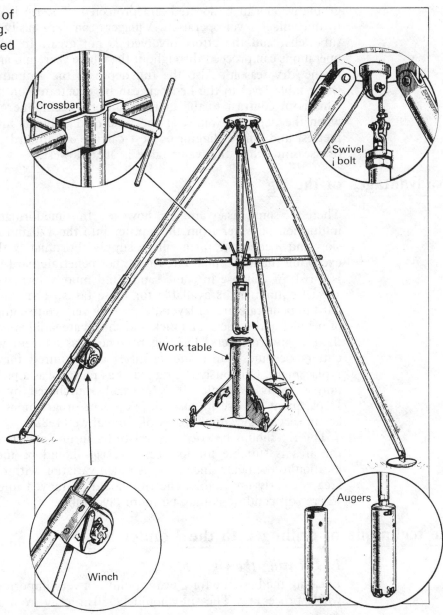

Working parts of the Vonder Rig. A hand-operated drilling rig.

Performance

The Vonder Rig drills 170 mm diameter holes. The time to unload and assemble on site is approximately 30 minutes. The initial drilling rate in average soil conditions should be 3.0 metres per hour. After 2-3 metres depth the drilling rate is 1.5 m/h. In harder formations 0.5 m/h. to 1.0 m/h. is more common. A 6 metre hole can be drilled in approximately 5 hours. A 12 metre hole can normally be drilled in 1.5 to 2 days.

Advantages of the rig

The obvious advantage of the rig is the speed and ease with which groundwater can be located and protected. The rig is perfectly suited to full village level operation. Villagers can very easily drill their own tubewells, and the effort involved is not great. In addition, drilling operations can proceed throughout the whole year, and are not restricted to the 'dry season'. Also the full depth of the groundwater from the water table level to the bedrock can be penetrated on most occasions. This is in contrast to the more traditional technique of well digging, when the depth of water, at the time of digging, is limited to 2 or 3 metres, and where digging is best carried out towards the end of the dry season, when the water table is at its lowest.

Disadvantages of the rig

There are some disadvantages, however. In some formations the rate of infiltration of water from the aquifer into the 170 mm drilling is very slow and inadequate for a village supply. Fortunately this is more the exception than the rule. The rig cannot penetrate hard rock, and there is a risk of collapse in some sandy and muddy formations. However, special equipment is available for these tasks. The augers also find it hard to penetrate gravel layers, especially very course gravels and layers of pebbles and stones. The pick and shovel are still better at penetrating these awkward layers. Drilling becomes less efficient when the auger cutters become blunt, and facilities are required for sharpening or replacement. Obviously the rig is not as portable as a pick and a shovel, nor as freely available. It is normally purchased by a Government Department for use in rural development programmes. Over 100 rigs are in service in various parts of Zimbabwe (1988).

The rig should be tried several times in one area to assess whether the area is suitable for its use. The rig should be moved from one location to the next, since there is much variation within an area. If the area is clearly unsuitable, the rig should be moved into another area, where soil conditions may be more favourable.

The technique of drilling with the Vonder Rig

1. Locating the right place

Locating the best site for a water point is a very important aspect of the whole procedure. The site should be 30 m away from a latrine, water

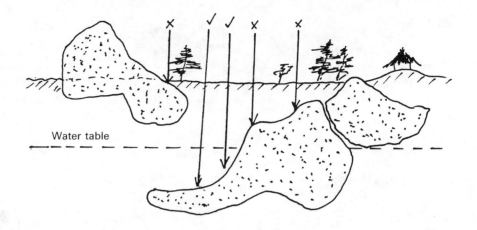

Some experimentation is required when drilling for water.

hole, cattle kraal and any other feature which might pollute the groundwater. It should be located in a slightly raised position so that run-off water from the protective apron runs away from the water point and does not drain back into it. The lie of the ground should be observed, and a note taken of the potential for rainwater catchment in the area. The services of a reputable water diviner may be invaluable to locate a high yielding groundwater stream. Health workers, community leaders and local villagers should also be consulted, since they are the people who will use the water point.

2. *Setting up the rig*

Once an exact site has been located it is important to set up the worktable exactly level. A plumb line is provided for this purpose. The plumb line is slung over the top of the worktable barrel, and the centre of the plumb should come to rest on the edge of the flange of the barrel. The line of the plumb string is indicated on the diagram of the rig. The plumb line is used on at least two sides of the worktable, thus ensuring that the table is laid exactly level.

The tripod is then screwed together (the legs come in two sections and are colour labelled), and erected directly over the worktable. In order to get an exact position for the tripod, the steel cable is lowered with the winch so that the stem linkage on the end of the cable falls exactly centrally within the barrel of the worktable.

The position of the worktable and the tripod can be adjusted slightly by chiselling away soil with a shovel from underneath the worktable or tripod respectively. The steel cable pulley and the threads for the drill stems should be oiled, and the worktable secured into position finally with the four steel pegs provided for the purpose.

3. *The drilling operation*

This is begun by attaching the crossbar to the first stem, so that it is about halfway down the stem. The stem is then attached to the stem linkage at the end of the steel cable. The crossbar and stem are then winched up and the auger attached through the bayonet adaptor. The

Level the worktable.

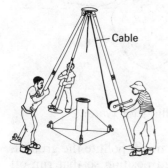
Position tripod accurately over worktable.

Oil all working parts.

Attach first stem to swivel bolt.

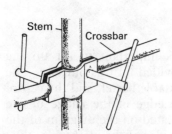
Attach the crossbar to the first stem and tighten.

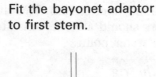
Fit the bayonet adaptor to first stem.

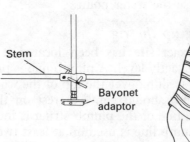
Insert the auger bar through the holes in the auger and place on the worktable.

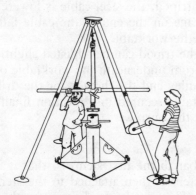
Fit the bayonet adaptor to auger.

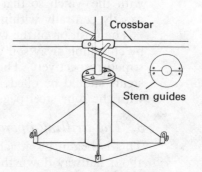
Fit stem guides around stem on worktable.

50

The crossbar is lowered into a convenient position and the crossbar bolts are tightened. The crossbar is turned in a clockwise direction to drill.

auger is then lowered through the barrel of the worktable and the two halves of the stem guide fitted over the studs at the top of the worktable. The auger is lowered further until its meets the ground, and the crossbar adjusted so that it lies about 300 mm above the guide. The actual drilling now commences.

The rig should be operated by at least five persons. One more experienced operator should take charge of the winch and be in overall charge of the rig. Four extra persons are required to turn the crossbar in a clockwise direction. As the crossbar is turned, pressure is exerted downwards. The cable supporting the stem is loosened with the winch, and this allows the stems and auger to move downwards, cutting into the soil beneath. In soft soils, the auger is able to penetrate about 300 mm in five or six turns of the stem. The stem can be seen to move downwards as the auger penetrates the ground, but stops moving downwards when the auger is filled with soil.

At this stage the stem and auger are winched out of the ground. This entails taking off the stem guides from the top of the worktable, and placing them temporarily on the pins fitted to the sides of the worktable. The auger is raised so that it protrudes slightly from the end of the worktable. A steel rod, provided with the rig, is now pushed through the holes in the upper end of the auger, and the bayonet fitting detached from the auger. The steel rod now holds the auger up inside the barrel of the worktable. The rod is used to remove the auger which is emptied by carefully knocking the open (upper) end on to a log of wood. In some cases the cuttings must be removed by rodding them out from the blade side of the auger.

Remove the auger after it has been filled with soil.

Once emptied, the rod is fitted back through the two holes in the top of the auger which is placed into the barrel of the worktable. The cable is then lowered and the bayonet adaptor attached to the auger. The auger is lowered and the stem guides refitted. The crossbar is retightened so that it lies about 300 mm above the stem guide. The cable is loosened, and the crossbar turned once again in a clockwise direction. A further auger full of cuttings is drilled out, and raised to the surface. This process of drilling, bringing the cuttings to the surface and redrilling is repeated. The winch is used to raise the stem and auger, and also to lower it back in position.

When one stems' depth has been drilled, a further stem is fitted, and the process is repeated. Obviously the more stems that are used, the greater the time it takes to go through one drilling cycle. If conditions are right, the drilling cycle can be repeated until bedrock is met. If harder conditions are met, the normal auger is replaced by the hole saw, and the process repeated.

Overcoming problems

Not all holes drilled by the rig are successful at first. In granite areas, it is possible that the auger may meet hard rock before it meets the water table. In this case it is best to change position and try again. In many cases a change in position of about 5 or 10 metres may enable the auger to penetrate much more deeply into the ground. The aim, of course, is to penetrate the groundwater table as deeply as possible.

If possible at least 3 metres of water should be contained in the base of the drilling at the driest time of the year, 5 metres depth of water is much better. The aim should be to get as much water as possible. Even 10 metres is not too much. Once the water table has been passed, drilling should always continue until bedrock is met.

In harder ground, it may be necessary for some people to sit on the crossbar to give the auger more 'bite' and speed up the drilling process. One or two people on either side of the crossbar is enough. If the ground is too hard, and the rig is loaded with too many people, the cutters will bend, wear out rapidly or break.

The drilling rig cannot cope with every situation. If bedrock lies in the ground above the level of the water table, only a big commercial drilling rig will be able to cope. It is important to establish the best areas in which the rig can operate. By going to these areas first, the drilling teams can gain their experience and confidence in easier ground. There is nothing like a few successful drillings to keep up the enthusiasm, not only of the drilling team, but also the villagers who participate. Sometimes the auger is left at the bottom of the hole if the stems are turned in an anti-clockwise direction by mistake. The auger can be relocated fairly easily by sending down the stems with the bayonet adaptor fitted, and carefully rotating the stems until the bayonet locks in position again on the auger.

Drilling holes through sand

The problem of penetrating through mobile mud and sand formations has been overcome by a development of V & W Engineering in which a steel casing is allowed to penetrate the formation at the same time as the auger. The steel casing actually precedes the auger in this case. This additional piece of equipment includes a low worktable. The worktable is fitted with a clamp which secures a special length of steel casing used to support the stem guides, when the normal auger is used. The same worktable also supports the specialised casing used to drill holes. When the steel casing itself is used for drilling, the worktable clamp is loosened, and an adaptor fitted to the top of the steel casing, which enables it to be turned with a crossbar. The first steel casing is equipped with a cutting edge similar to the hole saw.

The drilling procedure follows the normal pattern as with the standard rig. using the standard worktable and augers as described in the previous section. This drilling procedure continues until a difficult formation like sand or loose mud is encountered. At this stage the standard worktable is removed and the low-level worktable is fitted in its place (see illustration below).

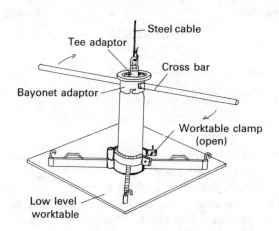 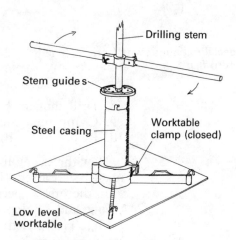

Drilling with steel casing.   Drilling with auger.

The standard auger and bayonet adaptor are also removed. A specialised bayonet adaptor is now fitted to the first steel casing which has a series of cutting teeth at its lower end for drilling. The swivel bolt attached to the steel cable is attached to the tee-shaped adaptor which allows the crossbar to pass through it. This tee adaptor is lowered inside the bayonet adaptor, and a crossbar passed through both the bayonet and tee adaptors (see illustration overleaf). This enables the bayonet adaptor and the steel casings beneath it to be turned.

The sequence is as follows:

1. Lower the swivel bolt into the bayonet adaptor.
2. Pass the crossbar through the openings in the bayonet adaptor.
3. Attach the first steel casing fitted with teeth to the bayonet.
4. Use the winch to raise the casing and crossbar.
5. Lower casing through worktable, and clamp tight.
6. Fit more lengths of casing until the sand layer is met.
7. Use the crossbar to turn the casing into the loose sand or mud.

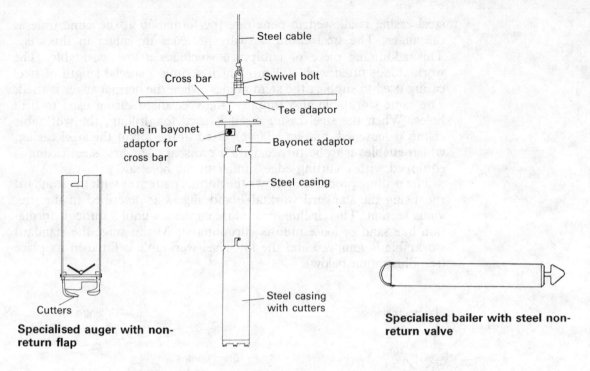

Specialised auger with non-return flap

Specialised bailer with steel non-return valve

8. Drill though the sand layer as far as possible.
9. Clamp the casing.
10. Remove crossbar and tee adaptor.

Now the casing lies in the soft layer of sand or mud and the contents must be removed. Two methods can be used.

In the first a specialised auger, narrow enough to fit within the steel casing and equipped with non-return flaps, is lowered through the casing. This penetrates loose material, like mud, which passes through the flaps, but when pulled up the flaps close off. Material like mud can be raised to the surface.

Bailing operation

The following steps should be taken when excavating through sand using a bailer.

1. Winch operator slackens cable.
2. Bailer operator activates cable by a sudden pull which allows the bailer to drop sharply.
3. This is repeated until the operator feels or hears that the bailer has filled. The bailer gets heavier as it fills.
4. It is important that the hands are protected by gloves or a piece of cardboard.

Note: It is important that the sand is loosened first within the steel casing before the bailer can operate. This is best achieved by first loosening the sand with a small 125 mm diameter auger which will pass down inside the steel casing. The sand will rise more easily into the bailer once it has been loosened.

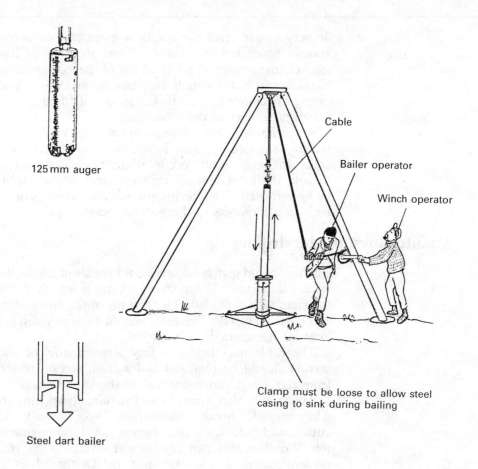

125 mm auger

Cable

Bailer operator

Winch operator

Clamp must be loose to allow steel casing to sink during bailing

Steel dart bailer

A specialised bailer is used and is operated on the end of the cable. The bailer is allowed to move up and down inside the steel casing so that it strikes the sandy surface. On its downward movement the bailer valve opens and allows sand to enter the bottom of the bailer when the sand is in a mobile state. As soon as the bailer is lifted the valve closes off and the sand become stiffer. The process is repeated, with the nett result that sand enters the bailer and cannot fall out. When the bailer feels full it is raised to the surface on the cable and emptied.

Once the sand or mud has been removed from inside the casing, the attachments are changed so that the casing itself can be turned again and lowered more deeply into the layer. The process of extracting the formation is repeated with the tool most suited for the material. This process is repeated until the full depth of the excavation is reached, preferably down to bedrock.

The next stage involves fitting the PVC casing within the steel casing. The size of the casing will depend on the pump to be fitted, and is normally 110 mm class 6 or class 10 for handpumps and 125 mm class 6 for the Bucket Pump. Before the casing is fitted an auger full of fine gravel or 6 mm granite chips is added to the drilling. Next the casing is fitted into the full length of the excavation, and a gravel packing is introduced around the PVC casing. This should be 6 mm granite chips or very coarse river sand with a grain size of over 2 mm. Once the annular space between the PVC and steel casing has been filled, the steel casing is turned and lifted out of the excavation with the crossbar. Extra lift from the winch may also be necessary. Each length of steel casing is removed. The PVC casing will then be surrounded by the gravel pack next to the formation.

A certain amount of experimentation will be required with this technique to penetrate the formation most effectively. The trouble with collapsing soils usually occurs under the water layer, and it is essential to hold the formation up with the steel casing whilst the cuttings are being removed. This technique is valuable in penetrating sandy river beds to gain access to the water beneath.

Maintenance of the drilling rig

The rig should give good service if kept clean and well oiled. Cleanliness is very important. When stems are not in use they should be stored on the stem rack provided. The threads of the stems should be kept clean and oiled regularly. The tools should be kept clean and the stem guide kept off the ground.

The cable may begin to fray after months of use and the frayed section should be removed and a fresh section of the cable unwound from the winch and connected to the stem linkage.

The cutters also wear out and require sharpening from time to time. They may also break under strain. V & W can re-sharpen and repair cutters and hole saws, but suggest that local engineering shops do the job. V & W supply new cutters with rivets, so that replacements can be made at the nearest workshop in the District. It is best to have spare sets of augers and hole saws.

Lining the drilled hole (normal soils)

Once the hole has been drilled to the greatest depth possible, and several metres of water have been found, it is important to line and protect the drilling as soon as possible and before the drilling collapses. If a successful drilling is anticipated, PVC casings should be immediately available to place down the hole. In addition a good gravel pack should also be available.

If a handpump is to be fitted (e.g. Bush Pump), the casing size is normally 110 mm and the lower section (1.5–3 metres) should be slotted with 0.8 mm slots. Slotting is not required with the Bucket Pump.

Before the casing is added at least one auger full of granite chips should carefully be added down the drilling to form a foundation on which the casing can stand. Normally 3 metre lengths of casing are lowered down the hole. The sections of casing should be PVC cemented together and held for one minute before lowering. The casings should be lowered until the entire tubewell is lined with a half metre of casing protruding above ground level.

The gravel pack is added into the annular space between the casing and the drilling. This should be carefully chosen. Ideally special sand should be used, having a grain size between 2.0 mm and 3.0 mm. However, sand of this size is not commonly available. It is best to sieve out and thoroughly wash river sand from areas where it is common, and pack this sand in bags ready for use at the drilling site. The sand should have a large grain size. 6 mm granite chips can also be used. Builders' sand and pit sand are unsuitable because they contain a high content of fine material which may silt up the tubewell and also slow down the flow of water from the aquifer into the casing. Some sands may even stop the flow of water. Very coarse river sand is the best and should always be carefully chosen and washed.

The gravel pack is added so that it fills the annular space to within at least one metre of the surface and preferably two metres. This final space is filled with a concrete mixture, well rammed down to form a good grout which seals off the gravel pack from the surface. The sanitary seal is important for the full protection of the tubewell.

Every tubewell should not only be fitted with a suitable pump, but also be built with a wide sanitary apron and water run-off channel and seepage area.

Community participation

The one great advantage of the rig is that it makes full community participation possible at village level. There are many examples in Zimbabwe where the rig is operated fully under the control of villagers. This has an important influence on the success of the final installation.

Sanitary surveys of wells and tubewells

The quality of drinking water derived from wells and tubewells is very often dependent on the degree of sanitary protection of the facility itself, although there is also much seasonal variation. Seasonal variation is less, however, in wells and tubewells which are properly sited and adequately protected from surface contamination and have a good handpump.

Although latrines are usually listed as one of the main potential sources of contamination for wells and tubewells, together with cattle kraals, it is most likely that contaminated water which spills off poorly designed headworks of a well is equally to blame for poor quality water. Many wells, even those presumed to be semi-protected, cause swampy areas around them, and these areas are very much subject to contamination because of the high frequency of human contact. It is surprising how a latrine, which may be 20 metres away from a well, may be considered more of a hazard than the insanitary swamp which may develop off the end of a short run-off only 3 metres away from the well. Whilst it is true that the latrine might well be considered a hazard, especially if the ground consists of fractured rock, there are many other factors which can cause well or tubewell water to become contaminated.

The aim of all protected wells and tubewells is to provide water which has the highest possible quality, bearing in mind that it will rarely, if ever, be treated with disinfectant. Water which lies in the ground away from hollows such as latrine pits, refuse pits, low-lying water-logged areas etc., can generally be regarded as relatively free from contamination caused by pathogenic organisms.

If a site is well chosen and the well drilled or dug into ground which is elevated and thus away from potentially waterlogged areas during the rainy season, the water which penetrates into the gallery of the well should be pure enough to drink. However, when the well is used as a source of water, it obviously attracts a great deal of human contact, and it is this which is the potential source of contamination that must be protected against.

The points to observe in the siting of a well or tubewell are as follows:

1. It should be in an elevated place, so that during the rainy season water will run away from it rather than into it.
2. It should be at least 30 metres away from a latrine and uphill of the latrine.
3. It should be at least 30 metres away from a cattle kraal, and uphill of the kraal.

4. It should be sited well away from any depressed area in the ground, such as a hollow used for a rubbish pit, a hollow used for brick making or any other area where water might collect.

Well lining

1. It is important that the well or tubewell be lined, especially near the surface, but preferably right down to the base of the well.
2. In the case of a tubewell, this will normally be lined with PVC casing to the bottom. A gravel pack is normally fitted around the casing. The gravel pack should not come to the surface however, but the annular space between the excavation and the casing should be filled with a watertight seal (clay, concrete or bentonite) for at least one and preferably two or three metres under the surface. The seal (grout) stops water which may build up around the head of the well from penetrating the gravel pack and polluting the water deeper down.
3. In the case of a wide diameter well, this should preferably be lined with cemented bricks or cement caste *in situ* down to the bottom. Concrete rings can also be used and should be cemented together above the water layer to the surface. The annular space between the brickwork or rings and the wall of the well should be filled with well cuttings or well washed river sand lower down the well. Nearer the surface the annular space should be filled with clay or concrete to act as a seal.
4. This seal will stop water flowing into the body of the well from the surface, through channels which may develop at the head of a well which is heavily used.
5. The brickwork or concrete lining of a well and the PVC lining of a tubewell should extend between 150 mm and 300 mm above ground level, and the surrounding apron should be built around this.

Well apron

1. The 'apron' which surrounds the central well or tubewell lining should be made as wide as possible from strongly made reinforced concrete. Preferably it should be 3 metres in diameter, so that it contains splash water which can then run away.
2. All surfaces should be sloped towards the water run-off channel. The edges of the apron should be raised.
3. The apron should be well reinforced with steel wire, to avoid cracking. Contaminated water can pass through cracks in a poorly made apron and contaminate the well beaneath.

Well covers should be made in one piece and not split, as the split is rarely sealed correctly and contaminated water can run directly into the well.

Well water run-off channel

1. The water run-off channel should be as long as possible, at least 6 metres and preferably 10 metres. It should be made with strong reinforced concrete or bricks mortared together on a solid base.

2. Since sediments accumulate on the channel, this should be cleaned down regularly by the users.
3. The water disposal system should be a soakaway or seepage area planted with bananas, trees, sugar cane or vegetables. It can also drain into a sump in the middle of a vegetable garden, if this is well managed. The seepage area also requires a lot of local care and management.

Pumping gear

1. The pump should be inspected to ensure that the mounting is secure, and that water from the apron does not run back into the well or tubewell through the pump head.
2. Water delivered from the pump should spill on to the apron so that waste water can find its way to the disposal system. If the water is piped away from the apron as in some designs of Bush and Nsimbi pumps, the waste water should also be able to find its way to a safe point well away from the site of the well or tubewell.

General points

Apart from areas covering the apron and water run-off channel, all surrounding areas of the water point should be relatively dry and well drained. All waste water should drain into the disposal system, which should be designed to cope with the volume. Swampy areas on existing water points should be upgraded and fitted with new aprons and water run-off channels (headworks). In some cases, pumps fitted to steel borehole casings have caused considerable erosion around the borehole, so much so that all the earth around the area becomes eroded and turns into swamp. This water can percolate down the casing — polluting the water beneath.

Sanitary surveys of this type should be carried out during the constructional phases of new water points and also on old water points which may need improving as a result. Surveys should take place throughout the year, but especially during the rains, when most problems occur. They should always be undertaken if there is a suspicion that the supply may be defective, due to a bad taste or to an outbreak of waterborne disease in the area served by the water point. It is always important for knowledge of the potential problems of poor sanitary protection of a water point to be taught to local communities, so they themselves may be in a position to detect and improve a weakness in their own supply. The Health Assistant, Health Orderly and Community Development Worker in particular should be able to identify problems of this type.

Handpumps

It is a widely held view that handpumps, when installed on wells and boreholes, provide one of the simplest and least costly methods of supplying rural populations with clean water. Reciprocating handpumps were used in Roman times, often for fire fighting and pumping out water from the bilges of boats. The concepts of the perforated plunger valve and leather seal were already conceived by Agricola in the 16th century, and were widely used in Europe after this date. Pumps were made of wood at first, but metal-bodied pumps became more common in England in the 17th century and many were made from lead. Machined mass-produced handpumps were being produced for wide distribution in the mid 19th century, and in the early 20th century there may have been 3000 handpump manufacturers in the USA alone. For a period, interest in the handpump was reduced after the more widespread introduction of mechanised pumps, especially on boreholes. More recently however, the importance of handpumps has been revived, especially for use in Developing Countries, mainly as a result of the initiatives of the International Drinking Water Supply and Sanitation Decade.

Most handpumps in common use today use the same basic operating principles as those designed centuries ago, and this suggests that the basic concept is sound. When reciprocating handpumps were widely used in countries such as the USA, they were designed for use by families and fitted in the homestead. In this situation they provided potable water, reliably and cheaply, for many years. In the modern day situation, handpumps installed in Developing Countries are often used by large communities, and the recurrent problem of maintenance looms as a major obstacle to the wholesale success of rural water supply projects.

According to Webster's International Dictionary, a pump is any machine that raises a liquid. In modern times, pumps have been thought of as machines equipped with cylinders and pistons, and indeed this type, known as the reciprocating handpump, is the most commonly used. However if the wider definition is used, by far the most successful pump ever designed by Man is the bucket and windlass, since this is also a machine that raises a liquid. This simple model has been successfully used for raising water from wells for many thousands of years, and indeed, in Zimbabwe today, more water is raised by this method than any other. There are approximately 15,000 reciprocating handpumps in regular use in Zimbabwe (1988), most of them Bush Pumps, and over 50,000 bucket and windlasses. The former raise water of higher quality, but the latter are managed and maintained by local communities.

This factor of maintenance is now known to be of the greatest significance. A reciprocating handpump, poorly maintained, ends up as scrap and of little use to the community it was intended to serve. For this reason 'upgraded wells' and a modernised bucket and windlass, the 'Bucket Pump', which are maintainable at village level have been endorsed by the Zimbabwe Government as viable technical options in the current rural water supply programme.

Types of handpump

Many hundreds of handpumps are now available on the international market, and many new designs have emerged directly as a result of the international water decade. There are many ways of classifying these into groups including the following:

1. Bucket type pumps
2. Direct action reciprocating pumps
3. Lever action reciprocating pumps
4. Rotary pumps

The first category is the most common throughout the developing world. It is very traditional, easily maintained, but very often delivers water of low quality.

Of the reciprocating pump types, the direct action pumps may be the most useful for shallow wells, but are currently less common than the lever operated pumps which can lift water from much greater depths, and thus are more versatile. In recent years PVC has been used more and more to replace steel as an engineering material. PVC does not corrode and can easily be cut and joined, and thus lends itself to simple maintenance procedures. It can however easily be broken, the inherent strength deteriorating over the years. Many more recently designed direct action pumps use a high content of PVC. The most rugged handpumps are still lever action types.

Rotary handpumps are dominated by the Mono Pump, of which there are two basic hand-driven types, the direct drive and the gearbox drive. The direct drive unit is made in Zimbabwe.

Description of the pump types

1. Bucket type pumps

This is often not considered as a pump at all by some organisations, and has for too long been ignored as a meaningful method of upgrading water supplies in Developing Countries. However, the evidence in favour of its use is very considerable, and there are positive signs that it is increasingly acceptable as a tool for upgrading poorly protected supplies.

Most bucket type pumps use a bucket and rope, or a bucket, rope and windlass. A combination of bucket, chain and windlass is the most hygienic. In some countries the bucket is pulled with a pulley wheel, which provides a mechanical advantage but may have little hygienic value. The 'shaduf' is a bucket and rope attached to a counterbalanced pole, and one of the earliest hygienic methods of raising water which dates back to 2500 BC. It is still used today in the Middle East. Elsewhere in Africa and the Middle East, the windlass is widely used as a means of raising the bucket and rope. This offers a mechanical advantage, but has the additional advantage of hygienically storing the

The bucket and windlass.

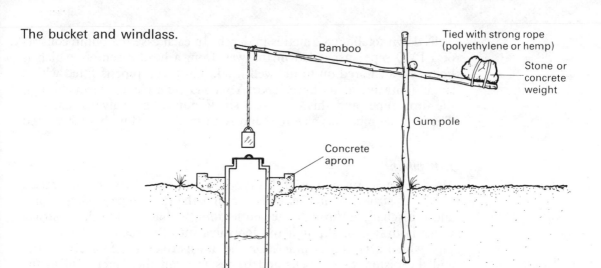

The shaduf: a successful method of lifting water from shallow wells.

chain or rope and preventing its contamination on the ground.

Bucket and windlass mechanisms are very common in Central Africa and there are tens of thousands operating successfully in Zimbabwe alone. However, most are placed on wide diameter wells which are poorly protected in other ways and have inadequate sanitary seals and headworks.

Contamination of the well can take place through many routes, notably through the chain and base of the bucket, if these make contact with the ground. When there is no raised collar, waste water can run back into the well, and during the rainy season this possibility increases. When the well is lined, fitted with a raised collar, a lid, surrounding concrete apron and water run-off, water quality can be enhanced considerably (see 'upgraded well'), and this makes the concept of a bucket and windlass more acceptable as a technical option.

This process has been taken one step further in Zimbabwe, where the concept has been modernised in the form of the 'Bucket Pump'. Primarily using the same techniques found in traditional practice, this pump employs a tubular bucket with a valve at its base. This passes up and down a static PVC casing, which is best placed in a tubewell. The bucket is raised and lowered within the casing and delivers water at a rate which varies between 5 and 10 litres per minute. The singular advantage of this system is that the water produced is substantially better in bacteriological quality than that taken out of most wide diameter wells.

2. *Direct action reciprocating pumps*

These are becoming more common in many rural programmes in several parts of the world. They include the Tara Pump (Bangladesh), Mark V (Malawi), Madzi (Malawi), Blair (Zimbabwe), Nsimbi (Zimbabwe), Rower (Bangladesh), Waterloo (IDREC), Wavin (Netherlands), Nira AF 85 (Finland), Ethiopia BP50 (Ethiopia), and many others.

Most are made with PVC components below ground level and steel components above ground level and are used exclusively on shallow wells down to about 12 metres in depth. In each case the pump consists of a PVC pipe suspended in the well from a head assembly which is bolted or anchored on to the well head. This drop pipe is fitted with a single footvalve at its lower end. A second pipe passes down through the drop pipe, and this too has a single non-return valve at its base. The internal pipe acts as a rod, and is normally fitted with a tee-shaped handle.

Lift pumps

Most direct action pumps of this type are lift pumps, water being raised on the upstroke. Water ascends through the drop pipe (rising main), and is discharged through a tee outlet into the bucket. On the upstroke water is drawn in through the footvalve into the section of the drop pipe which acts as a cylinder. On the downstroke, the footvalve closes and the piston valve opens and passes through the water held in the cylinder. On the upstroke the piston valve closes and water is raised above this and discharged at the outlet again. Many direct action pumps using the lift principle have hollow pushrods which are buoyant, which makes pumping easier. Neoprene seals are often used in combination with the PVC pipe. The water delivery rate usually lies between 15 and 30 litres per minute.

Blair Pump.   Nsimbi Pump.

Force pumps

One type of direct action pump uses the force pump principle — the Blair Pump and its derivatives made in Malawi, Papua New Guinea and the Philippines. In this case the piston valve is linked to the pump rod, which is hollow and acts as a water spout and handle combined. On the upstroke water is drawn into the lower end of the cylinder through the footvalve. On the downstroke, the footvalve closes and water is forced through the open piston valve, through the hollow pushrod to the surface and out of the 'walking stick' handle. This configuration uses less parts than other direct action pumps and is therefore simpler and cheaper to make. The delivery rate varies between 15 and 20 litres per minute.

3. Lever acting reciprocating pumps

These are the most common handpumps and very large numbers have been manufactured over the years. The long list includes the India Mark II (India), Maldev (Malawi), Afridev (Kenya), Nira AF76 (Finland), SWN 80/81 and Volanta (Holland), Petro (Sweden), Abi Vergnet (Ivory Coast), Swedpump (Sweden), Consallen and Climax (United Kingdom), National (RSA) and Bush Pump (Zimbabwe).

These pumps use steel head assemblies which support lever mechanisms. The type of lever mechanism varies considerably, but generally is based on a steel member rotating on roller or ball bearings. More recent bearings are made of hardwearing plastics (polyacetal in the Afridev). The most successful bearing used in the lever mechanism is hardwood. The Bush Pump is the only model listed above which uses this, and employs a teak block as a lever and bearing surface. This combination has made the Bush Pump one of the most successful handpumps used in the Developing World.

In almost all cases pump rods are made of mild or stainless steel, although fibreglass rods are used in the USA. The drop pipes (rising mains) are made of galvanised or black iron or PVC. Polyethylene is used in some cases. Most cylinders are made of brass, although some are made of stainless steel. Both 50 mm and 75 mm cylinders are used. Lever acting pumps are certainly more complex than direct action pumps, and thus are generally more difficult to maintain. They are more robust however, and are essential for all deep well and heavy duty settings. The Zimbabwe Bush Pump has been in service for over fifty years and is still the handpump of choice in Zimbabwe (1989). There are very few lever acting handpumps that could compete with it in terms of simplicity and endurance.

Bush Pump.

4. Rotary pumps

These are less familiar in rural development schemes but are being used more, as their low maintenance requirements are becoming more well known. They are simple and elegant in concept, but in practice their delivery rate is low, and may actually be more difficult to maintain than more conventional reciprocating pumps, although this will happen less frequently. The Monolift (United Kingdom) is becoming more commonly used in some countries together with the Moyno IV (USA) and the Direct Action Mono Handpump (RSA & Zimbabwe).

All these pumps use a rotor/stator system, with a steel helical rotor which turns within a helical rubber or elastomerical stator. The pump rods rotate. The delivery rate for handpumps varies from 8–16 litres per minute. The rotor system is robust and may operate for many years without attention. Motorised mono pumps are commonly used on boreholes and may operate for ten years or more without servicing.

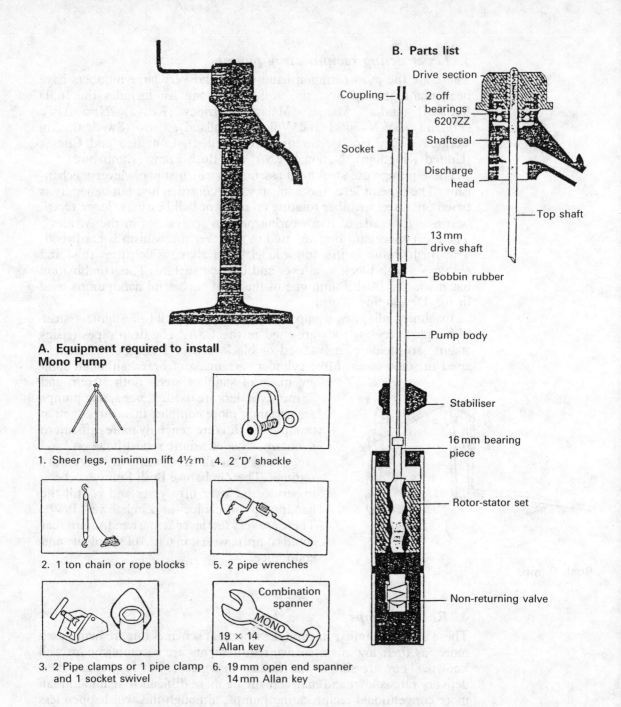

The mono direct drive handpump.

Handpumps used in Zimbabwe

The handpump was chosen many decades ago as the most practical method of delivering potable water to community water points in the rural areas of Zimbabwe. This method of raising water from the huge underground reservoir still applies today. Water tapped from adequately protected wells and boreholes is generally fit for drinking without the need for artificial treatment, and this is seen as a big advantage. However, the proportion of water sources which are adquately protected is low. The results of two studies (1983 Demographic Socio-Economic Study and 1984 National Socio-Economic Study) show that 35% of households in the communal lands have access to protected sources of water during the wet season and 42% during the dry season. In brief, two thirds of the households do not have all year round access to a safe water supply. Wells are the most common source of water and about 75% of these are used throughout the year, although few are adequately protected. Boreholes are the most common improved sources of water in the communal lands, and many of these are used by large numbers of households. Since about 25% of wells dry up every year, the reliance on existing wells drops during the dry season, whilst the reliance on boreholes increases.

Although at least ten different types of handpump have been used in Zimbabwe, only four of these, the Bucket Pump, Blair Pump, Nsimb Pump and Bush Pump, have been used in large numbers.

Zimbabwe's most successful handpump, known as the Bush Pump, has been used successfully for over fifty years in the rural areas, and remains the pump of choice for all heavily used community water points. All four pumps have been adapted for the range of conditions normally found in the communal lands, and are known to operate well when built and installed correctly, and when a suitable maintenance system is established. However, without maintenance the pumps can fail and remain out of order for months. It is therefore the maintenance system, rather than the pump itself which determines whether a handpump programme will be successful in the long term, assuming, of course, that technical faults in the pump itself have been reduced as far as possible.

With the basic consideration of maintenance in mind, the National Action Committee of the Zimbabwe Government made a recommendation in 1987 to standardise the handpumps used in the rural programme. Only two options were chosen by the committee, the Bush Pump and the Bucket Pump, with all other options being phased out of the programme for community settings. It was further recommended that just one design of Bush Pump be used in future, being a hybrid which combined the best features of a number of earlier Bush Pumps.

Specific situations were recommended for the use of these two pump options. The Bush Pump is to be used in all deep and heavy duty settings where the number of users is greater than ten families (60 persons). The Bucket pump is to be used in shallow and light duty settings where the number of users is less than ten families (60 persons). With the Bucket Pump, full village participation is expected in maintenance.

In addition, the committee recommended that the upgrading of traditional and family wells be encouraged, especially where these were perennial and in regular use. These were regarded as an important national resource which could no longer be ignored. Future programmes may include a large proportion of family based Upgraded wells with a correspondingly smaller proportion of community based Bucket Pumps.

The logic of the decision is clear enough. In the first instance, Bush

Pumps are well known, have a successful history in heavy duty settings and a maintenance system which is already established for them (by the DDF). Bucket Pumps can be managed by the user communities with little assistance from outside and when properly placed, they can deliver water of an acceptable quality. Whilst their delivery rate is low, they serve small communities well and maintenance costs are low. Traditional wells are common, and already represent a considerable national investment in labour. They are used daily by a considerable proportion of the total rural population, and thus the upgrading of these resources was considered vital.

Regarding the PVC bodied pumps, the Blair and the Nsimbi, these have both been installed in relatively large numbers (about 1000 of each), but without a national programme established to maintain them. In addition, whilst both are relatively simple in design and low in cost, there is little if any available evidence to show that local management by the user community has been successful without considerable assistance, both technical and in the supply of spare parts, from outside. In addition both pumps have a high content of PVC and a limited lifespan of perhaps five years in a communal setting. The committee felt that large investments for hardware should be made in buying pumps which were durable and where the 'backbone life' could be considered in decades rather than in years. In this regard, many Bush Pumps are known to provide service for twenty or thirty years and represent a good capital investment. PVC pumps are best placed in family or extended family settings, where they can provide good service for many years and thus represent a good investment.

This section provides detailed information on all four types of handpump still in use in Zimbabwe, namely the Bucket, Blair, Nsimbi and Bush Pumps. In addition pumps of all these types can be hand fabricated in local workshops in most parts of the world, and a description of hand built models is included.

Table 5. Recommended settings for 'handpumps'.

PUMP SYSTEM	DEPTH RANGE	MAX NO. USERS	IDEAL SETTING
Bucket/Windlass on well	1 – 15 metre	20	Family
Bucket Pump	1 – 15 metre	60	Small community
Bush Pump	1 – 100 metre	250	Large community
Blair Pump	1 – 12 metre	60	Family/small community
Nsimbi Pump	1 – 12 metre	60	Family/small community

NOTE: The bucket and windlass, Bucket Pump and Bush Pump are used in Government sponsored programmes. The Bush Pump is used most.

The Bucket Pump

The 'Bucket Pump' was first designed by Blair Research Laboratory in 1983 and first mass produced in 1984 by V & W Engineering, Harare. It underwent operational and bacteriological trials in Epworth between 1983 and 1986, and maintenance trials in the Gutu District of Masvingo Province between 1985 and 1987. In 1987 the Zimbabwe Government National Action Committee listed it as one of the two pump options to be recommended for use in the Zimbabwe Rural Water Supply Programme, the other option being a standardised Bush Pump. By early 1988 about 1500 units were in service throughout the country.

The Bucket Pump is a modernised version of the traditional bucket and windlass system and was designed specifically to simplify and reduce the costs of maintenance compared with reciprocating pumps, whilst maintaining an acceptable standard of drinking water quality.

By using principles which are well known in traditional practice, it is far more understandable than the reciprocating handpump, and consequently much more easily managed by rural communities. In Zimbabwe, it is the only pump where spontaneous maintenance and repair by village communities has been proven.

The Bucket Pump valve is a modernised version of the traditional bucket and windlass system and can raise high quality water, especially from tubewells.

Description

The pump consists of three basic units:

1. Pump stand including footings and windlass
2. Bucket and chain
3. PVC casing

In the mass-produced commercial unit the windlass has a wide barrel and is supported by two hardwood bearings mounted on a steel stand. The footings of the stand are mounted in concrete over the well or tubewell. The steel bucket is cylindrical in shape, equipped with a simple non-return valve at its base, and holds 5 litres of water. It is connected to the windlass through a 15 metre length of 4 mm chain. The bucket is raised and lowered through a length of 125 mm class 6 PVC casing which is either mounted in a hand-drilled tubewell or a wide diameter well. In most situations the pump is fitted to wells and tubewells down to a maximum depth of 15 metres. Water can be delivered either by tipping up the 5 litre bucket, or by placing it on the water discharge unit, the latter method being more hygienic.

The original hand-made unit can be made with wooden poles as windlass supports, and a steel windlass made to fit the specific well. The bucket can be made of steel tube with a valve made from a rubber disc. Alternatively the bucket can be made of PVC tube with a standard brass non-return valve. Hand-made buckets, however, do not have the life of the commercially made units.

The Bucket Pump: Designed for village level maintenance.

The Bucket Pump.

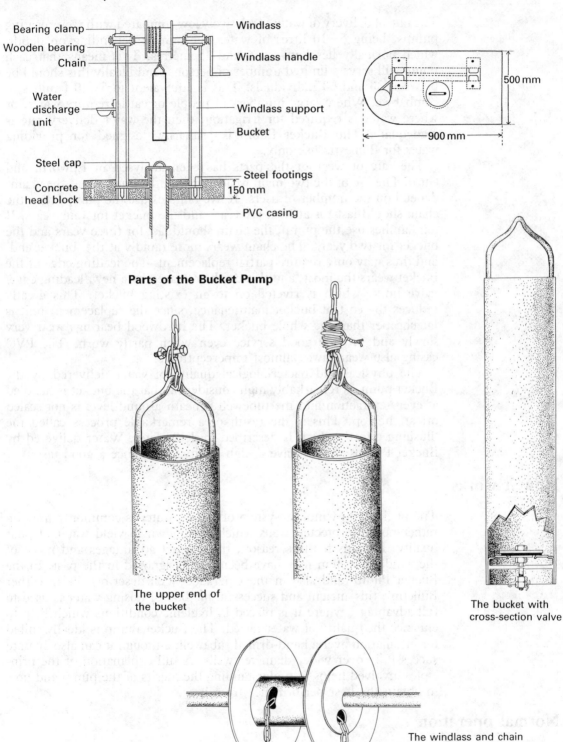

Parts of the Bucket Pump

The upper end of the bucket

The bucket with cross-section valve

The windlass and chain

71

Characteristics

The rate of delivery of water is relatively low compared with reciprocating pumps, being 5–10 litres of water per minute depending on depth, which normally lies between 6 and 15 metres. This means that each pump will serve a limited number of persons, and ideally this should be between 30 and 60 individuals. This is equivalent to 5–10 families in Zimbabwe. Where large numbers of people or cattle require water, or where water is required for irrigation, then the water delivery rate is inadequate. The Bucket Pump is primarily designed for providing water for domestic use only.

The rate of wear of the parts has been analysed in Epworth and Gutu. The life of the two main wearing parts, the bucket and the chain, depend on the number of users. If five families use the pump, then the chain should last for at least six years and the bucket for four years. If ten families use the pump, the chain should last for three years and the bucket for two years. The chain wears more rapidly at the 'bucket end' and thus may only require partial replacement. The leading edge of the bucket wears the most, and this can be replaced by a new 'leading edge/valve unit', which is riveted on to an existing bucket. This greatly reduces the cost of bucket maintenance, since the replacement unit is far cheaper than the whole bucket. The hardwood bearings wear very slowly and provide good service even when partly worn. The PVC casing also wears away almost imperceptibly.

The physical and bacteriological quality of water delivered by the Bucket pump is remarkably high considering that the bucket is handled at every extraction and the tubewell beneath ground level is not sealed off at the top. This is the result of a remarkable process called the 'flushing effect', which is described in detail later. Water delivered by Bucket Pumps should have a high clarity, and hence a good taste.

How it works

The bucket and windlass system of raising water is commonly used on rather poorly protected wells which are known to yield water of poor quality. Perhaps for this reason, the distinct advantages and merit of the windlass system may have been largely ignored in the past. In the Bucket Pump (and also in the Upgraded Well described in an earlier Bulletin), this ancient and successful method of raising water is used to full advantage, where it is placed in hygienic conditions which help to enhance the quality of water raised. The Bucket Pump is ideally suited for installation over a hand-drilled tubewell, although it can also operate successfully over wider diameter wells. A full explanation of the principles involved helps in understanding the merits of the pump and how to obtain the best performance from it.

Normal operation

When fitted to a hand-drilled tubewell, the system can be visualised as a series of tubes. The outer-most tube is the drilling itself which penetrates the water-bearing aquifer. Lying within the drilling is the

PVC casing, which is surrounded by a gravel packing. This allows water to percolate from the aquifer into the casing. The innermost tube is the bucket itself, which picks up water from the casing and brings it to the surface. A mono-directional movement of water takes place from the aquifer to the surface via the casing and bucket. There is a negligible return of water from the surface to the casing, and from the casing into the aquifer and this mechanism improves water quality considerably.

In the case of a wide diameter well, water enters the casing through a small hole drilled in the casing half a metre above the bottom of the well. In this case there is a slight return of water from the casing into the well chamber during the cycle, as the descending bucket strikes the water. The quality of water delivered from the pump can only be as good as the water held in the well chamber itself, and where an old well is upgraded with a Bucket Pump, it is best to clean out the well and preferably disinfect with chlorine before the pump is fitted.

The flushing effect

One of the most interesting features of the Bucket Pump is its ability to produce water of relatively high clarity and bacteriological quality.

The mechanism in tubewells

Whilst it is in use, water is constantly in motion through the different parts of the Bucket Pump. Every time water is extracted with the bucket the water level in the tubewell falls, and the difference in pressure head forces fresh water to pass from the aquifer, through the gravel packing, into the base of the casing, which is not slotted. Thus there is a high rate of turnover of water in the casing itself. For normal rates of extraction, the total content of the tubewell is completely changed within ten minutes.

Let us consider the case for a tubewell with 3 metres of water. The diameter of the drilling is 170 mm and the diameter of the casing is 125 mm. The bucket holds 5 litres of water.

The volume of water held in the casing is about 34 litres (internal diameter 120 mm) and the volume of water held in the gravel pack is about 16 litres (assuming 50% volume of gravel pack taken up by water). This amounts to a total of 50 litres. If each bucket removes 5 litres of water, the water in the tubewell will be replaced completely by fresh aquifer water for every 10 buckets of water extracted at the surface. This is a high rate of turnover of water (ratio 10:1). The rate of turnover will be increased if the tubewell contains less water or decreased if the tubewell contains more water. The ratio in a wide diameter well with the same depth of water and a 10 litre bucket would be about 300:1.

This high rate of change of water in the casing, where fresh water is constantly entering the base of the casing and being withdrawn from the top, has a significant effect on the final purity of the water. Buckets are handled at the surface, and might be expected to take down contaminants into the tubewell. However these are diluted rapidly as water is flushed through the system, fresh water entering the base of the tubewell to replace water being extracted at the surface. The mechanism is best

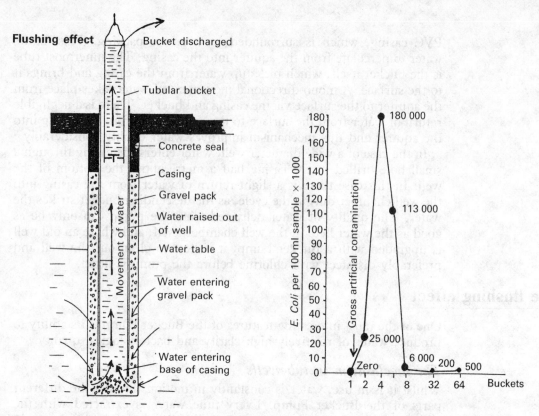

Bucket Pump: Flushing effect and dilution of bacteria.

described as a 'flushing effect', and is illustrated in the diagrams above.

Supporting evidence for the 'flushing effect' is shown in the figure which illustrates the results of an experiment where contaminants were added to a tubewell fitted with a Bucket Pump. In the experiment a sample was taken before the contamination was added and analysed for bacteria (Faecal *E. coli*/100 ml sample). The initial sample showed zero bacteria, but reached a peak in the second bucket withdrawn after contamination of 180,000 *E. coli*/100 ml sample. This was dramatically reduced in the following samples taken from buckets, showing how fresh water coming from beneath diluted the concentration of bacteria drawn up in the buckets.

The mechanism in wide diameter wells

The same flushing effect takes place in wide diameter wells if the installation has been made correctly. It is important that the mono-directional flow of water through the casing be maintained in order to achieve the high quality water expected from the Bucket Pump. Ideally water should not pass from the casing into the well chamber at all, but only from the well chamber to the well casing.

Formerly it was suggested that the casing of a Bucket Pump fitted to a well be slotted or drilled with small holes and the lower 0.75 metre length of casing filled with gravel. The gravel was intended to reduce the reverse flow of water. However many installations fitted to wells were not made in this way and were made by cutting the casing with slots only.

An improved technique has been developed in which a single 8 mm diameter hole is drilled in the PVC casing 0.5 metres above the base of the well. This allows water to enter the casing slowly from the well chamber after withdrawal of water to the surface, but limits the reverse flow of water from the casing back into the well chamber. This has the desired effect of reducing the level of contamination in the well chamber itself. In fact, the reverse flow through the hole only takes place when the bucket strikes the water when the water levels in the casing and the chamber are the same. This will occur after the pump has rested for a few minutes. After the first bucket withdrawal, in a new cycle of use, the water level in the casing lies below the standing water level in the well chamber and water flows rapidly through the hole into the casing to equalise pressure. When this flow is taking place, the impact of the bucket on the water is not sufficient to cause a reversal of flow. The result of this mechanism is that very little contamination can pass from the casing into the well chamber.

Other factors influencing water quality

Additional factors which enhance the quality of Bucket Pump water include the storage of the bucket itself in the casing, where it is less likely to be contaminated, and also the effective cap covering the head of the tubewell. In addition, the water discharge unit, fitted to commercially available pumps, reduces the potential contamination of the bucket by hands.

The Bucket Pump is fitted with a water discharge unit to reduce physical contact with the bucket.

There may also be advantages in using the tubewell as compared with the wider diameter well. Tubewells are easier to seal off at the surface, and this may help overcome short circuiting of contaminated water passing from the apron and its surroundings back into the well. Tubewells also extract water from a limited area of the aquifer and thus the area of potential contamination is reduced. When older wide diameter wells are fitted with a Bucket Pump, it is important to ensure that the well is thoroughly cleaned out and preferably sterilised with chlorine.

Bacteriological studies

A number of bacteriological studies have been carried out with a variety of pumps by Blair Research Laboratory. During 1984/85 water samples were taken from a series of traditional wells, Blair Pumps and Bucket Pumps in the Epworth area, close to Harare and analysed for faecal *E. coli* per 100 ml. sample. These results are tabulated in full in Table 6 and show that under ideal conditions water extracted from Bucket Pumps is only slightly more contaminated with bacteria than water raised with a reciprocating handpump (Blair Pump). Compared with water raised from poorly protected wells, the water is remarkably low in bacteria counts. This feature, combined with its ease of maintenance, makes the Bucket Pump ideal for use by small village communities.

Fitting the Bucket Pump to a tubewell

The Bucket Pump is ideally suited for installation on a hand-drilled tubewell, which usually penetrates a far greater depth of water in the ground than wider diameter wells. The installation of the Bucket Pump on a tubewell takes place in the following stages:

1. Locating the site

This should be in a raised site and at least 30 m away from latrines, cattle kraals and hollows in the ground and other sources of potential contamination. The location should be within easy access by the future users of the installation and should actually be chosen by them in consultation with the Health Assistant. On site, a water diviner may be able to detect the precise location where drilling may be most successful.

Table 6. Bacteriological data for groundwater.

Data collected for faecal *E. coli* per 100 ml sample from traditional wells and tubewells fitted with Blair Pumps and Bucket Pumps. Data collected from Epworth and analysed by the Blair Research Laboratory, Harare, using the multiple-tube/MacConkey Broth technique.

Date	Traditional wells								Bucket Pumps											Handpumps							Comment
	W57/	W58/	W59/	W61/	W62/	W63/	W64/		B9/	B10/	B11/	B13/	B16/	B17/	B18/	B19/	B20/	B21/	B23/	W3/	PP8/	W30/	W31/	W34/	W35/	W36/	
9. 1.84	65	140	550	350	1800	1600	1800		0	2	0	35	0	0	25	2	—	—	—	8	7	2	0	0	0	—	Heavy rain
16. 1.84	50	250	250	350	550	350	1800		8	225	2	0	0	0	9	550	—	—	—	275	45	0	0	7	17	—	
25. 1.84	20	25	550	1600	1800	225	225		0	5	0	0	0	2	0	7	—	—	—	5	20	0	0	0	0	—	
30. 1.84	1600	425	170	900	1800	95	35		25	70	0	4	0	0	110	2	—	—	—	5	0	0	0	7	2	—	Heavy rain
13. 2.84	35	110	225	95	1800	170	40		11	50	2	0	5	0	2	—	5	—	—	8	8	0	11	0	0	0	
20. 2.84	250	17	20	250	1600	900	350		0	0	0	0	2	2	0	2	5	—	—	5	0	17	0	0	5	2	
28. 2.84	50	1800	95	45	250	225	1600		0	2	0	0	8	2	5	2	5	—	—	13	0	0	0	5	0	0	
5. 3.84	130	550	80	550	550	350	80		0	0	14	0	8	17	7	2	0	—	—	5	14	0	2	0	2	0	
12. 3.84	350	350	550	1600	1600	350	350		0	7	5	0	0	7	14	11	5	—	—	5	50	2	0	0	2	0	
20. 3.84	250	40	425	550	225	250	1600		11	35	11	0	0	17	17	5	5	—	—	5	0	0	2	0	0	0	
26. 3.84	1600	17	250	550	225	170	120		0	11	8	0	2	5	0	2	2	0	—	2	5	2	9	11	0	0	
2. 4.84	550	1600	250	900	95	1800	1800		1600	2	2	5	5	2	5	350	0	350	—	250	2	5	0	5	2	0	Rains and flood
9. 4.84	225	35	14	40	140	1800	1800		0	4	0	2	4	5	7	2	0	14	—	0	0	DRY	0	2	5	0	
24. 4.84	50	2	6	7	11	50	80		0	0	0	4	4	2	17	0	0	0	0	5	0	0	0	0	0	0	
7. 5.84	170	8	40	1800	5	35	50		5	11	8	0	5	2	4	6	0	0	0	2	0	0	2	0	2	0	
14. 5.84	57	13	50	110	550	80	900		2	—	35	0	0	—	5	0	2	0	—	2	0	13	0	0	5	2	
11. 6.84	110	7	31	550	1800	14	140		0	0	0	0	0	2	0	DRY	—	DRY	0	0	0	0	0	0	2	0	
6. 8.84	2	0	5	0	0	0	DRY		0	0	9	8	2	0	0	—	0	0	4	0	130	0	5	0	11	36	Mid winter (no rain)
22. 8.84	25	250	25	2	110	70	—		0	0	0	2	0	2	0	—	0	0	5	7	0	0	2	0	5	2	
3. 9.84	2	4	4	4	0	12	—		0	0	0	0	0	0	0	0	0	0	0	0	0	0	0	0	0	0	
24. 9.84	130	14	—	20	17	32	—		0	0	2	0	0	0	0	0	2	0	0	4	0	0	0	0	0	0	
8.10.84	1800	4	25	8	2	—	—		8	0	0	0	2	0	DRY	—	17	0	0	0	0	0	0	0	2	2	
22.10.84	35	14	8	900	35	—	—		0	13	0	0	0	0	—	—	0	0	4	4	0	0	8	5	0	8	Rains
5.11.84	80	17	55	80	50	1800	—		5	4	0	0	0	2	0	—	5	0	5	5	0	13	5	8	0	2	Heavy rains
19.11.84	70	20	110	50	350	1800	—		2	0	5	0	0	5	0	—	5	0	0	50	0	2	7	0	8	0	Heavy rain
3.12.84	1800	70	1800	—	202	1800	—		2	50	8	0	2	2	2	—	2	0	0	13	25	0	5	5	2	2	Heavy flood
8. 1.85	350	40	110	1600	1800	900	—		0	0	0	0	80	2	17	—	7	0	2	25	17	8	25	2	2	14	Rains
21. 1.85	1600	1800	1800	1800	—	—	—		8	40	70	35	80	2	14	—	2	0	0	2	8	0	110	11	0	11	
11. 2.85	1800	55	110	35	1800	50	—		0	0	5	2	2	0	5	—	7	0	2	0	17	0	0	5	5	0	Rains
25. 2.85	350	11	275	20	225	130	—		4	0	0	0	0	2	0	—	2	0	0	2	0	2	0	0	0	25	Rains
11. 3.85	1800	35	550	170	1600	130	—		11	31	8	7	8	0	0	—	0	0	13	8	0	0	2	0	0	13	Heavy rain

Total *E. coli*	93653							Total *E. coli*	4358											Total *E. coli*	1466						
No. samples	197							No. samples	261											No. samples	191						
Mean *E. coli*	475.39							Mean *E. coli*	16.69											Mean *E. coli*	7.67						

NOTE: The unusually high *E. coli* count for B 10 on 2.4.84 was caused by a defect in the concrete apron which cracked, and also infiltration of contaminated water from a nearby hollow used for making bricks. These problems were corrected. They were all fitted to hand-drilled tubewells. Most Blair and Bucket Pumps used in this analysis were not commercially made units and had no water discharge unit attached. They were all fitted to hand-drilled tubewells. Most Blair and Bucket Pumps were less than two years old, most traditional wells were poorly protected and subject to contamination by surface water run-off.

2. Drilling the tubewell

After a suitable site has been located the hand operated drilling rig (Vonder Rig) is erected under the supervision of a Rig Operator. Villagers are encouraged to participate as much as possible in this activity. The drilling is continued until at least 3 metres of water are found. Every effort should be made to make the drilling as deep as possible until bedrock is met. Several tries may be necessary before an ideal tubewell is made.

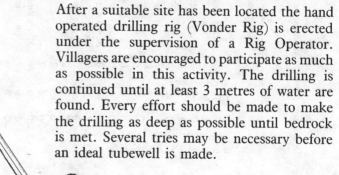

3. Fitting the casing

Once the hole has been drilled, the depth is measured and lengths of 125 mm class 6 PVC casing are prepared ready for placing down the drilling. Before the first casing is lowered a quantity of 6 mm granite chips (or equivalent local gravel) is gently poured into the tubewell to form a gravel bed at the base of the drilling. The volume required is the same as that which fills one earth auger used with the Rig.

Once this gravel has been placed in the tube, the casings can be carefully lowered one by one centrally into the tubewell. Ideally each length of casing should be PVC cemented to the next as it is lowered down the tubewell. This may not be necessary if the casings fit tightly together. Once the casings have been lowered down to the gravel bed at least 250 mm of the last casing should protrude above ground level.

Placing gravel in the drilling before the casing is added.

Adding the 125 mm class 6 PVC casing into the drilling.

4. Adding the gravel pack

This is an important part of the installation and should be carefully chosen. Ideally well-washed 6 mm gravel chips should be used for the gravel pack. However if care is taken in the selection, local gravel of a similar size can be used from a river. It is unwise to use sand, as this may contain fine material which will build up at the base of the casing and seriously slow down the rate of infiltration of water from the aquifer into the casing itself. If this happens the pump will appear to run out of water. The gravel is washed well and carefully added into the annular space between the casing and the drilling. Some plastic sheeting can be arranged on the ground around the tubewell head to avoid the gravel being contaminated with soil. Gravel is added until the level comes to within one metre of the surface. Clean water can be taken and flushed down through the gravel pack from the surface to ensure that it is well packed.

5. Adding the concrete seal

The final metre of annular space is filled with a strong mixture of concrete which can be made with 4 parts gravel, 2 parts river sand and 1 part cement. At least 250 mm of casing should protrude above the concrete layer. The Bucket Pump will be fitted on to this casing.

Adding the gravel pack to the annular space between drilling and casing.

Once the concrete seal has been added and levelled off, the casing is cut off about 250 mm above ground level.

Adding the concrete seal to the upper metre of the annular space.

6. Fitting the Bucket Pump to the tubewell

Before the pump is fitted to the tubewell ensure that the PVC casing has been cut off straight to the exact height required. This should be 230 mm above ground level. The uppermost part of the PVC casing fits within a steel tube in the base of the pump. It is important to mark and cut the pipe, so that the top is square with the ground. There is a small variation between pumps, so it is important to measure each one exactly.

Once the casing has been cut, the Bucket Pump can now be mounted directly over the tubewell, and arranged in the most suitable position. This will normally be arranged so that when the pump is being used, the user will face towards the water run-off channel. The pump should be adjusted so that it stands completely upright.

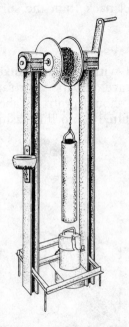

The new standard Bucket Pump is built with a set of steel footings which should be cast into a concrete block.
When fitted to a tubewell the block can be cast 'in situ' but is best made separately when fitted to a wide diameter well.

7. Casting the concrete foundation of the pump

The steel footings of the Bucket Pump must be mounted in a solid foundation of concrete.

This concrete is poured within a mould made of bricks laid around the base of the Bucket Pump. The bricks should be built up so that the final depth of concrete is 150 mm. This means that two layers of bricks will be needed. The concrete foundation should be about 900 mm long and 500 mm wide. It should shaped so that the foundation will support a 20 litre household water can, placed beneath the water discharge unit.

The mixture of concrete should be very strong containing 3 parts stone, 2 parts river sand and 1 part cement. This should be mixed well and added within the mould and around the footings of the pump so that the final depth is 150 mm (just at the lowest level of the cap chain). The concrete should be tapped so that it finds its way completely through the footings of the pump. It should be smoothed flat and left to cure.

The concrete block which forms the foundation of the Bucket Pump also serves as a bucket stand. The block is extended and rounded off on the side below the water discharge unit. The block is made 150 mm deep with a brick mould. Two courses of bricks are required. The concrete should be rammed well down and distributed around the pump footings. It should be levelled flat with a trowel when complete.

8. Making the apron and water run-off channel

Once the concrete foundation of the pump has been cast, a start can be made on the construction of the apron and water run-off channel. The apron should be at least 2 metres, and preferably 3 metres in diameter. Once the position has been marked on the ground, the bricks or stones can be laid down in a bed of concrete to form the rim of the apron and the side walls of the water run-off.

The rim of the apron and walls of the water run-off channel can be constructed whilst the concrete block is starting to cure. This may take several hours. Once this stage of construction is complete the bricks which form the mould of the concrete block can be removed and the area between the apron rim and the block is filled with reinforced concrete to form the apron itself. This should be sloped so that all waste water flows into the run-off channel. It should be smoothed with an iron float.

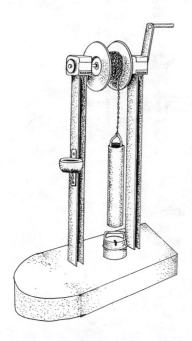

The water run-off channel should be constructed so that it allows water to flow downhill from the apron into an area of drainage. It should be at least 300 mm wide and at least 6 metres long. The preferred length is 10 metres. It should be made with two lines of bricks covered with a generous layer of concrete to form an open channel.

The apron itself should be made with a strong mixture of concrete laid between the apron rim and the pump foundation. A suitable mixture contains 3 parts stone, 2 parts river sand and 1 part cement. The bricks forming the mould of the pump foundation should first be removed and the concrete apron laid up to the foundation from the rim. The apron concrete should ideally be reinforced with lengths of 3 mm steel wire and shaped so that all the surfaces allow water to run into the water run-off channel. It should be built up to a depth of about 100 mm and the finish steel floated.

Both the apron and water channel can be made of reinforced concrete cast with a special steel mould which is available commercially (V & W Engineering). This makes the final structure neat and strong.

The concrete pump foundation and apron and water run-off should be neatened with a trowel and allowed to cure for several days and kept wet before being put into use.

The apron and soakaway.
These are built up with strong reinforced concrete. The outer rim of the apron and the sides of the water-run-off can be built with stones or bricks. All surfaces of the apron should slope towards the run-off, and the run-off should slope towards the seepage area.

9. The Soakaway

Waste water should not be allowed to lie about but should be absorbed by the soil in a good area of drainage. A soakaway filled with stones can be used, especially if the soil has good drainage properties. If drainage is more of a problem, bananas can be planted in a depressed area into which the waste water flows. As the bananas become more established they can absorb a lot of water.

10. Testing

The pump can be carefully tested while the apron is being made. After this the headworks should be left to cure for at least three days before the pump is put into full-time use.

11. Final checking

Before the pump is finally put to use, the valve in the bucket should be checked to see if it seats correctly and holds water. The chain should be adjusted in length so that it lies just above the gravel at the base of the tubewell in its lowest position.

All the nuts and bolts should be checked for tightness with the spanners provided with the pump. The wooden bearings should have been boiled and cooled in oil to act as a lubricant.

Once these operations have been completed, local leaders and villagers should be asked to attend a demonstration. The bucket is lowered down the tubewell and allowed to fill with water. A characteristic sound can be heard when the bucket is full. The bucket is raised and discharged at the water discharge unit over a bucket. The water will be turbid at first, but should soon clarify. It should be emphasised that the steel cap should be replaced after use, and that children should be discouraged from throwing items down the tubewell. The pump can be put into use.

Very often the villagers themselves will have been very active at every stage of the installation. Indeed there are many cases where the siting, drilling, installation and subsequent maintenance of this pump are carried out entirely by village communities.

Final Bucket Pump installation.

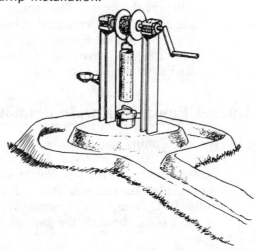

Fitting a Bucket Pump to a tubewell (Method 2)

This method is very similar to the first technique, but the Bucket Pump unit is added to the tubewell in a slightly different way, without first making a concrete head block around the pump footings.

This technique can be used when the apron and water run-off are cast in concrete within a special 2 metre diameter mould made for the purpose. Once the tubewell has been drilled to the maximum depth possible, and the casing and gravel packing have been added together with the concrete seal in the annular space, the casing is cut and the Bucket Pump is fitted over the casing directly. The steel footings of the pump are supported by the concrete which is poured into the mould to make the apron of the installation. The full installation can be carried out in the following stages.

1. Select suitable site.
2. Erect hand drilling rig and together with full community participation drill a tubewell to the maximum depth possible. This should hold at least 3 metres of water and preferably 6 metres of water in the dry season.
3. Select and wash the gravel packing which should be fine gravel or 6 mm granite chips.
4. Once the drilling is complete add one auger full of washed gravel chips carefully to the drilling.
5. Add the 125 mm class 6 casing length by length, cementing sections together, until the tubewell is completely cased with 300 mm of casing above ground level.
6. Fill the annular space between the drilling and the casing with washed gravel to within 1 metre of the surface.
7. Fill the remaining space with a strong mixture of concrete.
8. Measure the correct height to cut off the casing pipe, by referring to the Bucket Pump, and cut the casing off straight at this point.
9. Mount the Bucket Pump on the casing and make level.
10. Assemble the apron and run-off mould around the pump, so that the pump is central, and the run-off is facing in a downhill direction.
11. Make a strong concrete mixture 4 parts stone, 2 parts washed river sand and 1 part cement and add to mould. Build the base up to 150 mm thickness. The rim of the apron and the run-off channel are also made in concrete. About four bags of cement are required for this process. Ensure that all surfaces are well-shaped for water drainage and have a steel floated finish. Complete by building a soakaway for seepage area.

Fitting the Bucket Pump to a wide diameter well

Bucket Pumps can also be fitted to either newly-dug wells or to deepened existing wells with success. The higher quality of the water extracted from a Bucket Pump compared with a well fitted with a standard bucket and windlass is due to the high rate of change of water in the PVC casing of the pump. Water must move up from the bottom of the casing through the bucket to the surface in one direction. A

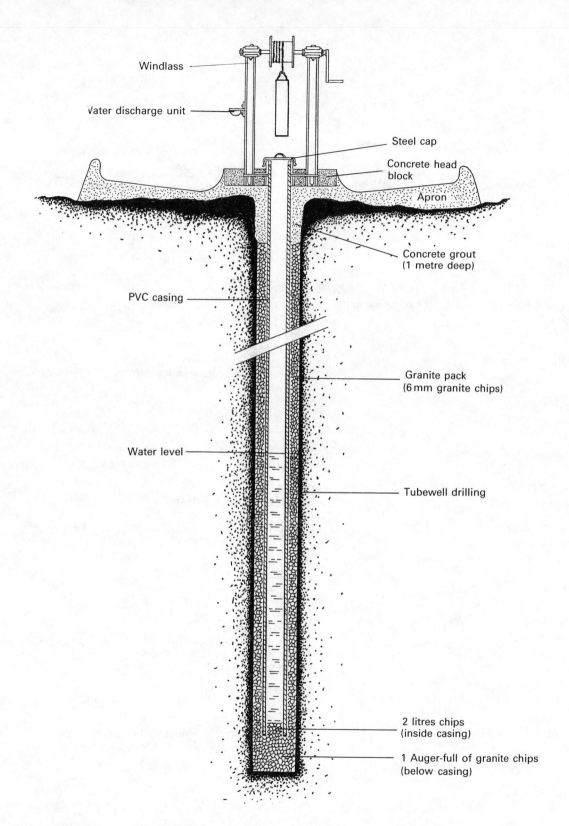

Bucket Pump fitted on a tubewell.

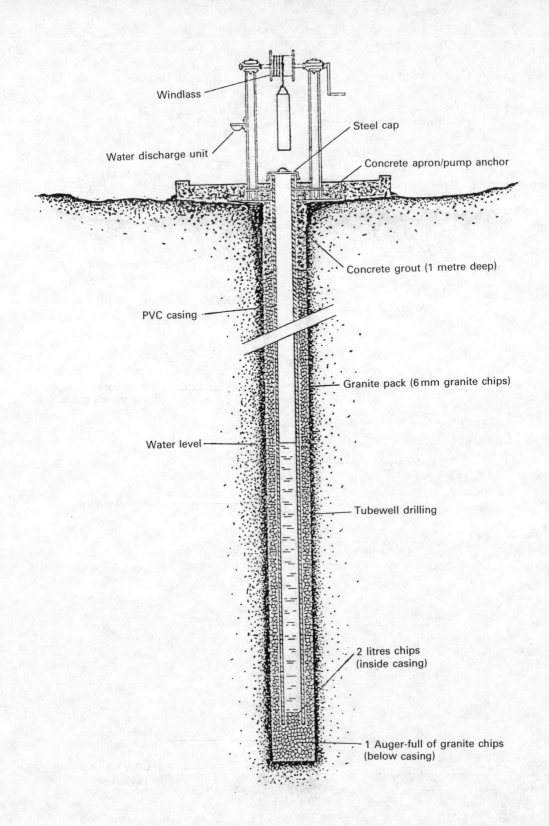

Bucket Pump fitted on a tubewell (Method 2).

flushing effect can therefore be guaranteed, which dilutes any contaminants coming from the surface. The return of water held in the casing back into the well must be reduced to a minimum. In the tubewell, the gravel pack acts as a buffer which reduces this return flow to a minimum. However when a Bucket Pump is fitted to a well the casing is not surrounded by any packing and some other means must be employed to reduce the reverse flow. This is achieved by drilling a single 8 mm hole in the casing 0.5 m above the bottom. Water can enter this hole fast enough to keep up with the removal of water in the bucket, but it does not allow much water to flow from the casing to the well chamber immediately following the impact of the bucket on the water. The installation of a Bucket Pump on a well takes place in the following stages:

1. Location of site

This is the same as for a tubewell, and should be at least 30 m from a potential form of contamination like a latrine, cattle kraal or deep hollow in the ground. The well should always be sited uphill of any potential source of contamination.

2. Well digging and lining

Existing wells should be deepened as much as possible and preferably during the months of October or November when the water is at its lowest level in the ground.

A considerable effort should be made to penetrate the ground as deeply as possible below the water table. The well should be lined with bricks, or concrete from top to bottom, and raised above the ground about 300 mm.

3. Casting the well slab

A reinforced concrete well slab should be cast near the well so that it covers the span of the well (which will normally be 1.2 m to 1.5 m). This should be made 75 mm thick with a mixture of concrete using 4 parts gravel, 2 parts river sand and one part cement. A 350 mm diameter hole is left in the middle of the slab, which is left to cure for at least 5 days, then washed and cemented in place at the well head.

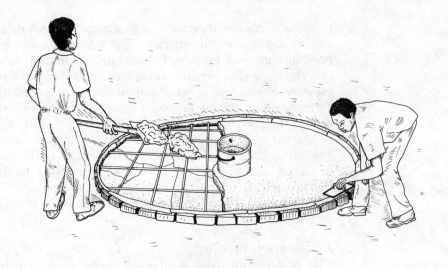

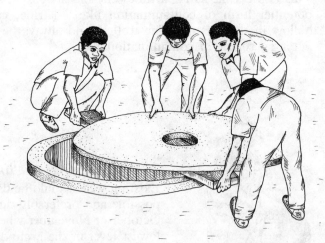

4. Concrete footing for casing

The 125 mm class 6 PVC casing which forms part of the Bucket Pump is supported by a concrete footing which rests on the bottom of the well. Each section of casing is delivered with one plain end and one socketed end. A 150 mm length of the socketed end of one of the lengths of PVC casing is cut off and embedded in a 300 mm diameter piece of concrete. This can be made by placing the socket in the middle of a plastic basin or circle of half bricks with the wider end uppermost. A strong mixture of concrete (3 parts gravel, 2 parts river sand and 1 part cement) is laid around the PVC to a depth of 75 mm. The concrete is also added within the socket to a depth of 50 mm. The socket must be made exactly level and should be left to cure for three days.

5. Preparing the PVC casing

A suitable number of lengths of 125 mm class 6 PVC casing should be supplied with each pump. PVC solvent cement is also required. The first length of PVC casing is cemented to the socket in the concrete footing. Next a single 8 mm hole is drilled neatly in the casing 500 mm up from the bottom of the footing. This hole can be made slightly larger (up to 10 mm) but not smaller.

Drilling the 8 mm hole in the PVC casing. This is drilled 500 mm above the bottom of the concrete footing.

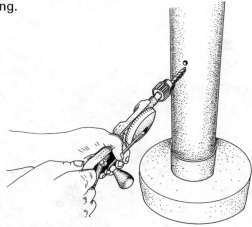

6. Lowering the casing

The casing with footing attached is lowered into the well. At least two people should hold it and slip it through their hands carefully. The next length is then cemented to the first length, held for one minute and then lowered again. Further lengths of casing are added to the required depth. An ideal maximum is 15 metres. A length of at least 250 mm should protrude above the level of the concrete well cover prior to final measurement and cutting.

7. Cutting the casing

The casing should be measured and cut off square to the required length. However it should be noted that on some wells, especially those with soft bottoms, the footing and its casing may sink a few centimetres into the mud before it finally settles. In this case, it is wise to proceed with making the apron and run-off before finally cutting the casing. The exact length of protruding casing will depend on the technique used to fit the pump. This is described in the next section.

8. Making and fitting the Bucket Pump head block

When a Bucket Pump is fitted to a well, the pump must be mounted in a concrete foundation which rests on the well cover slab. The well cover slab is made with a central 350 mm diameter hole, through which the concrete footing of the casing together with the full length of the casing is lowered. The pump head must therefore be mounted over this

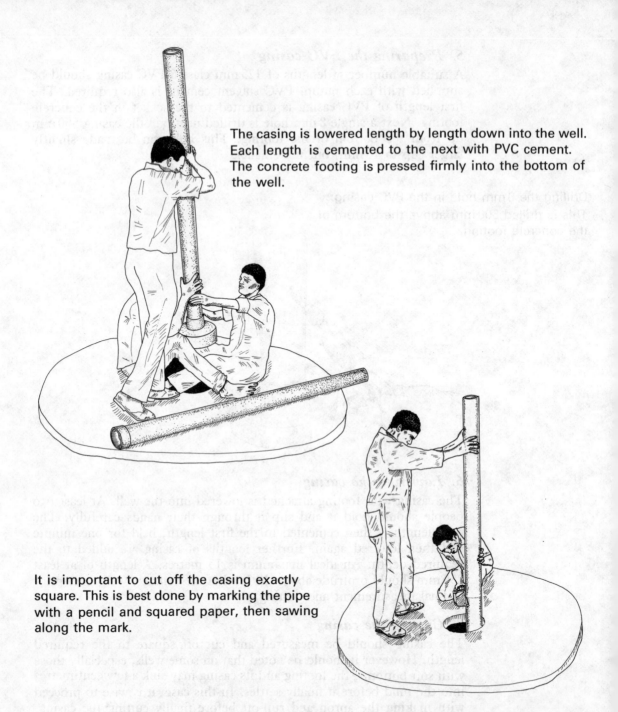

The casing is lowered length by length down into the well. Each length is cemented to the next with PVC cement. The concrete footing is pressed firmly into the bottom of the well.

It is important to cut off the casing exactly square. This is best done by marking the pipe with a pencil and squared paper, then sawing along the mark.

central hole. This is best achieved by casting the pump in a concrete block at the side of the well, and once it has cured, moving it on to the well cover.

The pump should be placed on a levelled piece of ground near the well, preferably on a sheet of plastic. The pump is arranged so that it is perfectly upright, and two layers of bricks are placed in position around its base to make a mould into which the concrete can be poured. Since the pump and its heavy concrete head block must be moved together on to the well after curing, the block should not be made too large. A

Stages in making the Bucket Pump concrete head block.

The standard Bucket Pump   Adding concrete mixture to the brick mould   The Bucket Pump with concrete head block complete

suitable size is 700 mm long × 500 mm wide × 150 mm deep. The mixture of concrete should contain 3 parts stone, 2 parts river sand and 1 part cement. This should be poured into the brick mould to surround the steel footings of the pump stand completely. It should be built up to a final thickness of 150 mm and smoothed flat and left to cure for several days and kept wet at all times.

Before the pump is moved into its final position, the casing should be cut off square at exactly the correct length. This may vary a little from pump to pump. Normally the PVC casing slides into the steel tube at the base of the pump head a distance of 230 mm. Allowing for 10 cm of cement mortar under the head block, the casing should be cut off 240 mm above the well slab.

A layer of cement mortar is then laid on the well slab around the central hole. The pump (and its concrete base) can now be moved into position. The pump and base are heavy and several people will be required to lift and fit them in position. The pump should be raised up and lowered down square on to the casing, and bedded into the cement mortar.

A brick extension of the concrete base of the pump should be made directly under the water discharge unit. Buckets are placed on this when they are filled with water. It is possible to cast a concrete pump base large enough to incorporate the bucket stand, but this makes the whole pump unit very difficult to handle in practice.

9. *Making the apron, water run-off and drainage area*

These are made in exactly the same way as for the tubewell, but in this case placed around the well cover. The concrete apron should be made

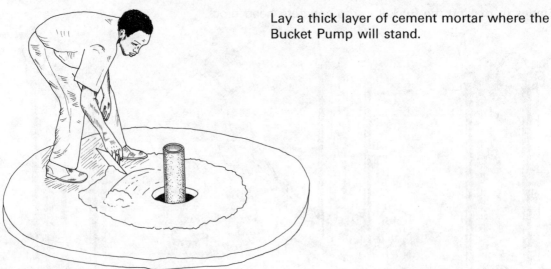

Lay a thick layer of cement mortar where the Bucket Pump will stand.

Take great care when lifting the complete Bucket Pump on to the well cover. After the block has cured, it should be tilted back and cleaned with water. Ensure that the casing is cut off at the correct length.

Lift the pump and base to an upright position — at first on to the well cover, and then over the casing.

Several strong people will be required to perform this movement.

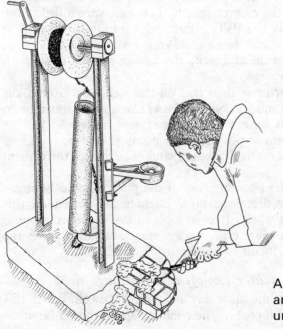

A stand for the bucket is made from bricks and mortar directly under the water discharge unit. The figure shows this half built.

of reinforced concrete and sloped down from the rim to the central well head. All surfaces should drain water into the water run-off and then to the drainage area. It is important to ensure that the well cover is sealed with mortar to the well collar. The entire headworks should be allowed to cure for several days and be kept wet at all times.

The drainage area can be a hole in the ground filled with rocks, or a depressed area filled with bananas or trees. Alternatively, waste water can be led into a vegetable garden.

11. Testing and final checking
This is performed in the same way as described earlier for the tubewell.

Fitting the Bucket Pump to a wide diameter well (Method 2)

This technique is useful where the existing water is deep or the chances of the well drying up in the future are minimal. This method makes the protection of a wide diameter well relatively easy and quite fast, but there will not be any chance of deepening the well in future without major excavation. The method is illustrated in the diagram below.

Bucket Pump fitted on standard well.

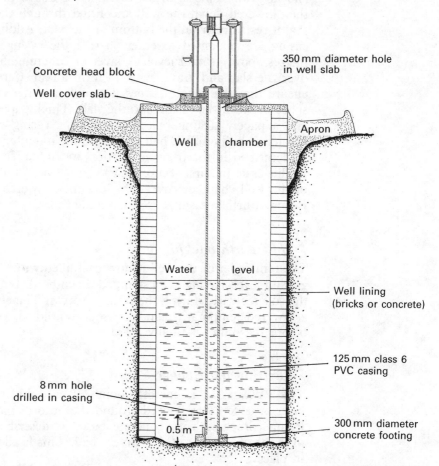

Technique

1. Well chamber lining

The well is deepened as far as possible in the driest part of the year when the water table is at its lowest. Next a circular wall of unmortared bricks, 1 metre in diameter, is built up about 2 or 3 metres from the bottom, depending on the depth of water. A space is left between the brickwork and the soil, and this is filled with gravel, so that it is level with the brickwork. The gravel packing stabilises the unmortared brick tube.

2. Well chamber slab

A strong reinforced concrete cover slab 75 mm thick, is made to sit on top of the brickwork well lining, with a central hole cast in it 130 mm in diameter. This will accept the 125 mm PVC casing fitted with a PVC end cap. When cured the slab is lowered on to the bricks.

3. Fitting the casing

A PVC end cap is cemented on to the end of the 125 mm PVC casing, and an 8 mm hole is drilled 0.5 metres above the cap. The casing is then lowered down the well and fitted through the hole in the slab so that it rests firmly on the bottom of the well. Additional lengths of PVC casing are cemented together so that the casing protrudes about 0.5 metres above ground level. A layer of concrete 75 mm thick is poured over the slab and gravel pack to form a seal. Care should be taken to ensure that the concrete does not pass into the gravel or through the space between the casing and the slab. Thick paper laid over the gravel and a paper packing placed around the casing will ensure this. For greater stability a PVC half socket can be mounted in a concrete footing as described in the previous section and mounted on the well bottom. In this case, the first section of PVC must be cut off just above the level of the well chamber cover which is then lowered in position over the casing. Further casings are then added.

4. The earth backfill

The remainder of the well is now backfilled with soil and levelled off at ground level, having been stamped down hard. It is best to stamp down the soil regularly during the filling process. The end of the PVC casing is covered temporarily, and the soil allowed to settle for several days, before the pump is finally fitted.

5. Fitting the Bucket Pump

The PVC casing is cut off about 230 mm above ground level, and a layer of concrete is poured around the head of the casing. The Bucket Pump (prepared with concrete base, as described earlier) is lowered over the casing or to the concrete bed. This is allowed to settle and set in place.

6. Apron and run-off channel

The apron and water run-off channel are built in strong concrete as described earlier. All surfaces should be sloped to allow all waste water to drain off the headworks into the seepage area, which should be at least 6 metres away from the well. Once all concrete work has set thoroughly, the pump can be put to use.

This method is valuable when a well can be dug deep into a reliable aquifer.

Wells upgraded in this way are well protected from surface contamination.

The well can only be deepened with great difficulty however, and this must be taken into consideration before the technique is used.

It saves on the cost of lining the well if a Bucket Pump is used and is a fast technique.

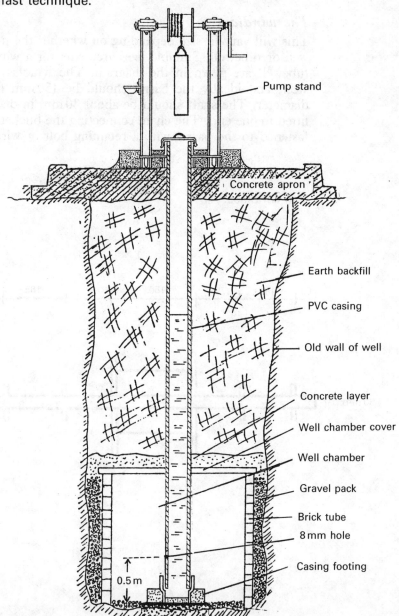

The hand-made Bucket Pump

The first Bucket Pumps were hand-fabricated and used wooden poles as windlass supports; this method is still practical for local production.

The pump stand

This can be made from gum poles or any suitable timber with a diameter of approximately 150 mm. These should be about 2 metres long and embedded into two holes dug on either side of the well or tubewell. The holes should be about 600 mm to 800 mm deep, thus leaving sufficient height of the windlass above the well or tubewell to give the bucket clearance. A 100 mm deep slot should be cut into each pole to accept the windlass. The slots should be oiled or greased before fitting the windlass.

The windlass

This will vary in size depending on whether the pump is made to fit a well or tubewell. Suitable measurements for a windlass designed for a tubewell are given in the diagram. The windlass should be made of steel, and ideally the barrel should be 150 mm long and 100 mm in diameter. The shaft should be about 30 mm in diameter with a handle fitted to one end. The chain connecting the bucket with the windlass is fastened to the barrel with a retaining bolt or wire.

The windlass.

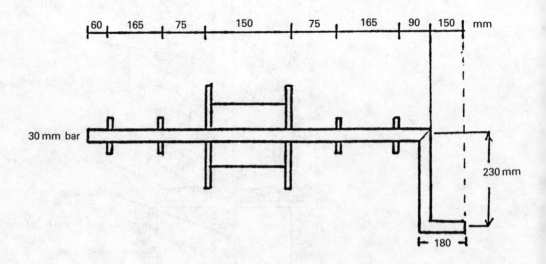

The steel tubewell head

The PVC casing should be protected at its upper end with a steel tube and fitted with a steel cap. This is best made with a 300 mm length of steel pipe that fits around the 125 mm casing. The steel pipe is concreted in position around the casing so that it projects about 200 mm above ground level. The PVC casing is cut off at the same level. The space between the casing and steel pipe can be filled with a very strong concrete mortar or epoxy resin. A steel cover or cap is made to fit over this pipe.

The bucket

Buckets can be made in steel by a tinsmith, but any strong tubular material will do. Buckets should be made tubular in shape with a handle at the upper end and a non-return valve at the lower end. They should hold about 5 litres of water. One method involves making the bucket from a 700 mm length of 110 mm class 10 PVC pipe. At the lower end a brass 20 mm non-return valve is packed in a core of hard-setting putty (Trinepon 6). The core containing the valve is made separately in a shorter length of PVC piping and glued into the PVC tube with hard-setting liquid adhesive (Trinepon 13). The handle can be made with a length of stout steel wire threaded through two holes drilled in the upper end of the PVC pipe. It is wise to thicken the wall of the bucket at this point by cementing a short length of PVC pipe inside the first and drilling through both pieces. The second piece should be 75 mm long and suitably cut so that it will fit inside the main pipe.

Commercially-made buckets are more durable, but hand-made units can last well if strongly made and carefully used. It should be remembered that no bucket will last for ever on a well. Some wells in the

Local tinsmiths also play an active role in the production of durable lids and buckets in rural water-supply programmes.

Middle East have been in continuous use for over 3000 years. The number of replacement buckets used on these wells probably exceeds 1000!

Final fitting

Once the gum poles and steel head have been embedded in concrete, the complete apron and water run-off are made as in other pump installations.

Stages in the installation of a home-built Bucket Pump on a tubewell

1. Drill the hole as deep as possible — with at least 3 metres of water in the base of the excavation.
2. Add 500 mm depth of 6 mm granite chips to the tubewell (do not bail the tubewell as this may cause the walls to collapse before the casing is inserted).
3. Measure the exact depth of the excavation and prepare the casing so that the steel head will protrude 200 mm above ground level.
4. Lower the casing into the excavation in stages, 3 metres at a time. Support the lower section whilst cementing the next section to it with a PVC socket. Lower the sections until the casing rests on the gravel at the base of the tubewell.
5. Fill the space between the drilling and the casing with 6 mm granite chips. The upper metre of this space should be filled with concrete. Cut off the casing 200 mm above ground level.
6. Insert a 300 mm long steel sleeve around the casing so that it is embedded in the concrete for 100 mm and protrudes above ground level for 200 mm. Carefully fill the space between the casing and steel sleeve with concrete.
7. With a hand auger drill two holes for the gum poles. These are about 600 mm to 800 mm deep depending on the length of the poles. After the holes are drilled, test them by inserting the two gum poles with the windlass loosely fitted. The clearance between the barrel of the windlass and the steel head of the casing should be sufficient for the bucket to be withdrawn without obstruction. The chain of the windlass should pass down the centre line of the casing.
8. Backfill around the gum poles with concrete and ensure that they are correctly positioned. Leave to set overnight.
9. By using bricks and concrete make a strong apron and water run-off channel around the pump. Build up strong concrete around the steel sleeve allowing sufficient space for the cap to be fitted. Allow to cure.
10. Grease the gum poles and fit the windlass, chain and bucket and adjust the chain length to suit the tubewell, so that the bucket does not hit the bottom.

Maintenance of the Bucket Pump

The Bucket Pump was specifically designed so that it could be maintained in a village setting by the user community. This gives the pump

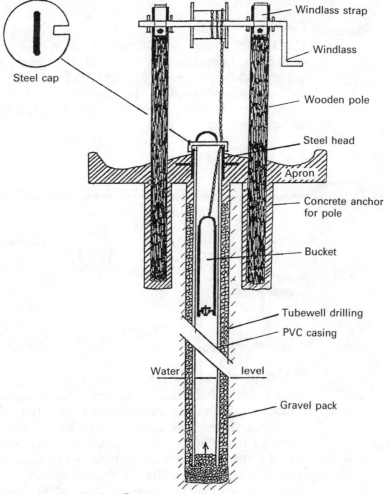

Hand made Bucket Pump.

characteristics which are not found in other pumps, which may require specialised servicing. Spontaneous management of the pump is possible at village level because the Bucket Pump is a modern version of the traditional bucket and windlass system which is familiar in most rural areas. The one specialised part, the bucket valve, is not found in traditional practice, but even this part is frequently repaired by village craftsmen. None of the parts is concealed from view, and even the single valve can be inspected each time the bucket is raised.

Simple as the Bucket Pump is, it still requires some understanding, by those who use it. Like all machines the Bucket Pump is subject to wear. The links of the chain will become thinner after a few years and the leading edge of the bucket also requires replacement periodically, as it wears back as a result of frequently hitting the water at great speed and being put down on the concrete apron. A new leading edge assembly for the bucket is available which includes the valve seat. The wooden blocks also wear in time, but all these parts are available as spares from the manufacturer of the Bucket pump. In some cases local stores have begun to stock a few spares for the pump.

Tools

The Bucket Pump is sold with a set of tools, which include a flat spanner for tightening the bearing blocks and the water discharge unit, a short socket spanner (double ended) for the nut on the valve and the nut and bolt attached to the chain, and a long socket spanner, for holding the valve in place within the bucket. The tools are illustrated below.

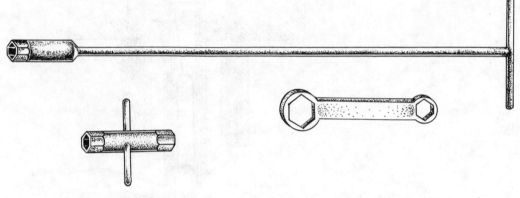

Bucket Pump tools.

Maintenance of the bucket valve

Most of the maintenance problems found in the Bucket Pump are associated with the bucket valve, which is the only specialised part of the pump. Even this, however, can be built in the village as many examples have shown. The original commercial valve has improved in design since 1984 when it first became available. The commercial valve is made with a rubber disc held in place with a washer against a steel disc. The stem of the valve is attached to the disc with a threaded section and a nut. The stem is held in the base of the bucket with another two nuts and a cotter pin. These nuts can come loose if they are not properly tightened, or sometimes even when they have been tightened. Occasionally the rubber washer can come loose and separate from the steel disc, which leads to a leaky valve. If a particular valve gives repeated problems it can be bonded together with a fast-setting liquid adhesive (Trinepon 6), or be replaced altogether with a new valve. Spare valves are cheap and the local Pump Caretaker or Head of the Village Water Committee should keep a small number of spare units. Spare valves should be available in the village and preferably stocked at the village store in areas where the pumps are common.

Local users of the Bucket Pump have devised several methods of either replacing the nuts or of replacing the entire valve if it falls apart. These include cutting rubber discs from car tyres or tubes and attaching them to a long nut and bolt. Since the lower part of the valve stem is drilled with a hole for the attachment of a pin, the lower nut can be replaced with a length of wire coiled around the stem and passed through the hole. In fact, in practice this is far more effective than the nut itself.

The nut and bolt fitted through the chain to attach it to the bucket

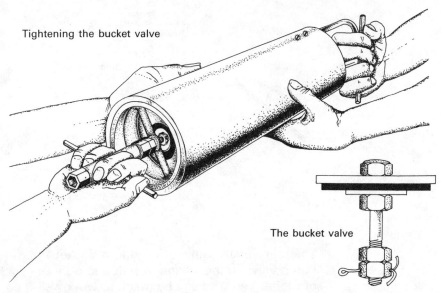

Details of the Bucket and the Bucket valve.

If the valve is lost or develops a fault villagers are able to repair and adapt the valve. An original valve is shown in the lower right hand corner (X). All other valves have been made or adapted in the village setting.

The Bucket Pump valve is the only non-indigenous part of the pump and is best retained in position with wire rather than by nuts.

can also come adrift. This too is also commonly replaced by a length of steel wire. In both these cases the wire is far more effective than the nut and bolt.

Leaky valves are not always caused by a fault in the valve itself. Sometimes, especially in newly drilled tubewells, bits of sand or grit can stir up and enter the valve, so that it will not close properly. This happens particularly when the chain is not tied off with wire at the right length and allows the bucket to reach the bottom of the tubewell. It is easy to clear the valve of grit and adjust the length of the chain so that the bucket cannot reach the base of the tubewell.

101

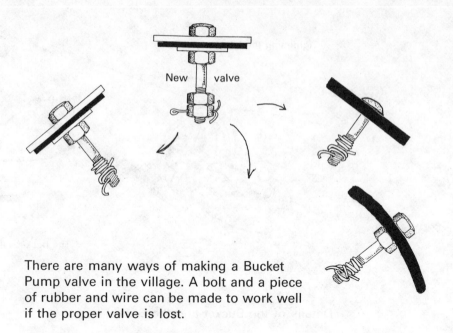

There are many ways of making a Bucket Pump valve in the village. A bolt and a piece of rubber and wire can be made to work well if the proper valve is lost.

Maintaining the bucket

The bucket itself is also subject to wear. The leading edge wears most and may wear away to the point where the bucket leaks after a few years of use. It is possible to replace the whole bucket, but the manufacturers supply a new bucket leading edge assembly including the valve seat, which can be fitted on to the main body of an older bucket.

Once the bucket is badly worn at the leading edge, the valve is removed and put to one side. The end of the bucket is then sawn off straight 25 mm behind the leading edge. The new leading edge/valve assembly is then tapped into the bucket so that it penetrates about 10 mm inside the bucket. The assembly is supplied with two rivets. Two holes are drilled on either side and through the walls of the bucket, so they also pass through the new assembly. The rivets are then pushed through the holes from the inside, and holding the head of the rivet firmly on a solid base, each rivet is flattened out on the outside of the bucket with a hammer. The protruding section of the bucket is then tapped over the leading edge to form a strong watertight seal. Finally the valve is checked and cleaned and fitted back on to the bucket.

Maintaining the chain

This part wears and can separate when the links become thinner. Broken links can easily be repaired by wiring the broken sections together. The chain wears far more on the bucket side than on the windlass side. When the chain is worn thin on the bucket side, it can be unwound from the windlass and re-wound the other way round. This will give the whole chain added life. Eventually the entire chain must be replaced with new chain. Other materials like rope or cable can be used if the chain is lost.

The chain is best wired to the bucket right from the start.

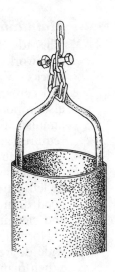

Nuts and bolts do not work well at this point.

A new leading edge/valve assembly can be fitted to an old bucket by a tinsmith or in a local workshop. The new leading edge costs far less then a complete bucket.

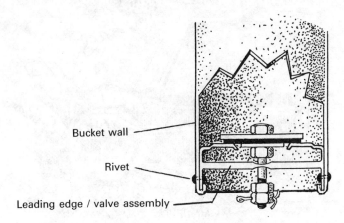

Bucket wall

Rivet

Leading edge / valve assembly

Nuts and bolts

Nuts and bolts can come loose and should be checked periodically and tightened. This is a simple task, but one often ignored. Special tools are provided with each pump, and these should be kept close at hand.

Maintaining the wooden blocks and housings

These should last for many years. Sometimes, however, the nuts holding the steel housing of the wooden block can come loose and this can accelerate the rate of wear of the blocks. When the blocks are very badly worn they should be replaced. Local timber can be shaped if necessary to fit within the housings.

Maintaining the headworks

The advantages of fitting a pump to a well can be reduced if the headworks (apron and water run-off) are poorly made and develop cracks. Waste and contaminated water can pass through a cracked slab and contaminate the well or tubewell very easily. With tubewells it is also important to make a good concrete seal in the upper metre of the annular space between the casing and the drilling. A good seal is also required around a well. It is important to make a wide, well-shaped and strong reinforced concrete apron which allows all the waste water to drain into a long well-made water run-off. When in use no pools should be left standing in the headworks and all waste water should pass into a good drainage area. The apron and run-off require constant attention and cleaning.

Normal soakaways fill up with sediment after constant use, and a wide depression in which bananas are planted may be more effective. When plants are young they are rather delicate and can die. However, with care they can grow into a big plantation and take out a great deal of waste water from the ground.

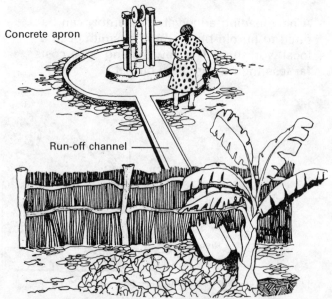

Keep the apron and water run-off clean.

Potential problems with the Bucket Pump

There is almost no mechanism that does not have problems of some sort, and the Bucket Pump is no exception. Some of these are listed below.

1. Bucket lost down the well

Sometimes the bucket or bucket and chain falls down the tubewell. It is a simple matter to fish these out with a long wire with a hooked end. This takes only a few minutes and the pump is working again.

2. Chain theft

Chains can also be stolen from the Bucket Pump, but this is not common in the rural areas. When a chain is stolen, it can be replaced by a number of local equivalents including washing lines, steel cable, electric cable and natural fibres. Chains can be replaced by local contribution or in some cases by the Ministry of Health. Occasionally the bucket vanishes with the chain. It is possible for a local tinsmith to make tubular buckets of this type and even fit them with valves. However, these buckets are not normally as durable as the commercially made units, but as the state of the art improves, so will the quality of the locally made buckets.

3. Objects thrown down the tube

Since the first pumps were installed in 1983, there have been very few cases of this type, even by children. This indicates that the high degree of community involvement with drilling the tubewell and pump installation, leaves the community with a sense of ownership.

However, two cases of deliberate fouling are on record. In one case a boy urinated down the tube and in the other a human stool was deposited in the bucket. In both these rather extreme cases the bucket was removed, cleaned and the system flushed by the local community. In the latter case, the community refused to use the facility until it was checked by Ministry of Health staff who flushed out the tubewell several times with the bucket and added a heavy dose of chlorine to the tube. After a few hours the unit was put back into use.

4. Loose steel pump stand

In earlier models fitted to wells, the steel footings were not adequate, and no specifications were made for fitting these to the well head. This resulted in several pump stands becoming loose on the well head. The Bucket Pump has now been standardised so that it can be fitted solidly either over a well or tubewell. The steel footings are designed for embedding in a concrete base before the pump is fitted over the well.

5. Poor gravel packing

In some cases the rate of extraction of water from the tubewell can exceed the recharge rate. This happens in soils which have a high content of clay. This is not always the case, however, and very often is caused by the use of the wrong gravel packing. If the annular space surrounding the casing is filled with fine sand, the flow of water from the aquifer into the casing will be considerably slowed down. Fine gravel (3–6 mm) is the best material for the gravel packing. The importance of a good gravel pack cannot be overemphasised, and may make the difference between a high yielding and low yielding tubewell. It is worth importing a good gravel from other areas in sacks or in a lorry if local conditions do not yield a good material.

6. Poor siting of the well

The final quality of the water produced by any pump may depend on the siting. In the case of the Bucket Pump, which is not a sealed unit, the final water quality may depend, in addition, on a high rate of turnover of water in the casing. The rate of turnover of water in the casing is reduced if the tubewell is almost full of water. Thus if the tubewell is sited in a vlei (wetland) with a very high water table, the potential for contamination of the groundwater from the surface is increased and the turnover rate of tubewell water is decreased. For this reason, Bucket Pumps placed in vlei areas may yield water of poorer quality compared with pumps fitted on higher ground.

Community participation

In Zimbabwe, village level participation is actively encouraged in all water and sanitation schemes. It is now well established that without this participation, communities cannot generate the commitment for maintenance as they do when they are involved.

The introduction of the hand operated drilling rig and the Bucket Pump brought with it an era when village communities could completely participate in the siting, excavation, installation and maintenance of their own protected water supply.

At the beginning of any new programme, communities are shown the technology itself by health workers who explain how the drilling rig and pump work. A general meeting is organised for each Ward where future development is planned. The main aim of these meetings is to inform villagers about the objectives of the programme, what role they are expected to play and how they are expected to organise themselves at Ward Village level. It is at these meetings that the importance of community participation is emphasised, and the levels of supervision required. Agreement is reached on the locations for the distribution of materials and hardware (pumps, casings, rigs, etc.) needed for the project.

Community participation is vital to the success of all rural water supply and sanitation programmes.

At this stage training workshops are held at Ward and Village level at which the technology is again described and demonstrated to the future users.

The community contributions are as follows:

1. Select site with assistance from Health Assistants.
2. Organise transport for moving drilling rig and pump.
3. Provide building materials like sand and stone.
4. Organise labour for drilling/digging and the completion of the head-works and in fitting the pump itself. Very often the whole task is performed by the community.
5. Subsequent maintenance of the pump.

Pump distribution

This may be an important factor to the future success of pump maintenance. In several projects, pumps are placed so that each one serves about 5 families (30 persons). This arrangement ties in with the extended family system in Zimbabwe. The families using a single installation are closely related and may already be accustomed to using their property collectively and sharing financial responsibilities. It is very possible that the distribution of pumps to suit the extended family system may be very crucial for successful village level maintenance.

Pump maintenance at village level

The installation stage of a rural water supply programme is very easily achieved compared with accomplishing sustained maintenance. The long term success of any water programme depends almost entirely on effective maintenance, and yet it is an aspect which is very often neglected.

The Bucket Pump was designed for villagers to maintain themselves.

It is frequently said that the most successful water projects must have a high degree of community participation in the area of maintenance. However, this is only possible if village communities can relate to the technologies used. This is possible with the Bucket Pump because it is a copy of a traditionally used system. This means that in practice, if the pump fails, the user community can quickly repair it using local knowledge and resources.

It has now been established that with limited training on valve repair or replacement, and often with no training at all, villagers are both capable and willing to maintain their own Bucket Pumps with limited support from the Ministry of Health who may provide, through Village Water Committees, the small number of spares required to keep the pump in action over a number of years. In most cases, repair of the pump consists of rewiring broken lengths of chain or tightening the valve. Where a spare part like a valve is required, this can be stocked by a local store, or held by the Village Water Committee, or Village Community Worker or Health Assistant. Valves can also be made locally from bolts and rubber.

A short lesson on how to fit or repair a valve is usually more than adequate. In practice the Bucket Pump should be much easier to repair than a bicycle.

Other areas of maintenance carried out by the community include sweeping aprons and keeping the water run-off clear. This is practical in units owned by a few families, but far less so in heavily used communal units, where there is no sense of ownership.

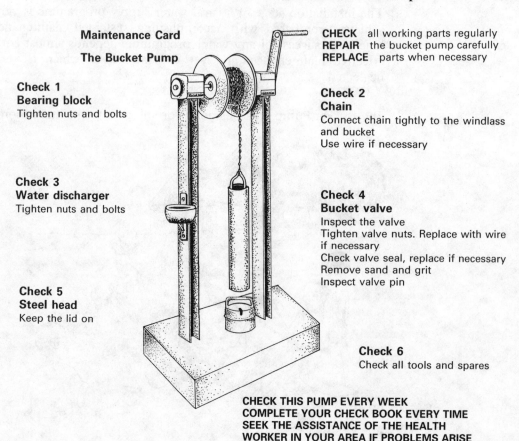

Maintenance Card

The Bucket Pump

CHECK all working parts regularly
REPAIR the bucket pump carefully
REPLACE parts when necessary

Check 1
Bearing block
Tighten nuts and bolts

Check 2
Chain
Connect chain tightly to the windlass and bucket
Use wire if necessary

Check 3
Water discharger
Tighten nuts and bolts

Check 4
Bucket valve
Inspect the valve
Tighten valve nuts. Replace with wire if necessary
Check valve seal, replace if necessary
Remove sand and grit
Inspect valve pin

Check 5
Steel head
Keep the lid on

Check 6
Check all tools and spares

CHECK THIS PUMP EVERY WEEK
COMPLETE YOUR CHECK BOOK EVERY TIME
SEEK THE ASSISTANCE OF THE HEALTH
WORKER IN YOUR AREA IF PROBLEMS ARISE

The Blair Pump

The original Blair Pump was designed by the Blair Research Laboratory in 1976. It was designed specifically for Ministry of Health operations because at that time no other simple handpump suitable for shallow wells existed. Other handpumps that might have been used on shallow wells were relatively complex in design and expensive by comparison.

Between 1976 and 1980 about 500 hand-made Blair Pumps were installed in Zimbabwe on shallow wells down to 6 metres in depth. During this era the pump evolved through the Mark Ia, Ib and Ic stages. These pumps were robust and intended for heavy duty use, but could only pump from a maximum of 6 metre depth. At the time, groundwater rose to exceptionally high levels in Zimbabwe, but has since lowered considerably. When the Blair Pump was mass-produced in 1980 it was designed to pump water from a depth of 12 metres, but was built from lighter materials, making it less durable and suitable for lighter duty settings only.

The Blair Pump was designed because low cost, easy to maintain handpumps were unavailable in Zimbabwe at the time, and were almost unknown internationally. Simple maintenance was a key factor which prompted the development of the pump, and even today, when many

Blair Pumps are ideally used by small communities or families — they are not intended for heavy duty use.

The new Blair Pump head block.

The Blair Pump is ideal for a family or extended family.

It operates on shallow wells down to 12 metres in depth.

other direct action pumps have been designed, the Blair Pump remains one of the simplest reciprocating handpumps ever designed. It is also one of the cheapest and simplest to repair and maintain.

Experience had shown that when the water level was less than 12 metres below ground level, neither the lever nor the watertight washers used on more conventional handpumps were necessary to pump water to the surface. The Blair Pump therefore dispensed with both the seal and the lever.

The Blair Pump can be made by hand from standard fittings and pipes, and large numbers have been made in this way. For this reason it was introduced into the Zimbabwe School Curriculum as a working example of a hand-operated pump. Although many Blair designed models have been mass-produced, it was always intended that the basic principles used in these designs could be adapted for use in many situations and could be hand-fabricated all over the world. Blair Pumps, bearing other names, have been built and are used in several other parts of the world.

Description of the Blair Pump

The Blair Pump is the simplest and one of the cheapest of all reciprocating handpumps. It uses a high content of PVC below ground level and steel above ground level. The pump consists of two pipes, each fitted with a non-return valve at its lower end. The outer pipe, which is slung from the pump head, acts as a cylinder. The inner pipe acts as both a pushrod and a spout — water is delivered to the surface through the handle. The lower end of the pushrod is expanded to form a piston enclosing a non-return valve. This piston forms a loose fit within the cylinder, and is deliberately leaky, even when new.

Blair Pumps were originally designed to be hand-made from standard fittings. The simplest is the low-cost family model made almost entirely from PVC. Blair Pumps form part of the 'O' level school syllabus in Zimbabwe.

The Blair Pump handle also acts as a water spout. This is the commercial model.

Mark I pumps

These were the earliest models and were designed with tough 90 mm class 16 cylinders and 20 mm galvanised steel pushrods. The piston valve is made from a standard 20 mm brass flap valve embedded in hard-setting putty and cast within a ring of 75 mm class 16 PVC pipe. The footvalve is a standard brass type.

Family model

The light duty family model is also designed for hand fabrication and is made almost entirely from PVC using a 40 mm class 10 pipe for the cylinder and 25 mm class 26 for the pushrod and handle. Valves are made from marbles seated on brass washers enclosed in a PVC shell. The lower footvalve is extractable through the cylinder.

Mass-produced model

In the Mark VI (mass-produced) model the cylinder is made from 50 mm class 16 PVC pipe. The PVC cylinder is epoxy bonded through an adaptor to the steel pump head, which is fabricated with a baseplate. This baseplate is bolted on to a baseplate support which is mounted in a concrete head block. The pushrod has a lower component of 25 mm class 26 PVC and an upper component of 20 mm galvanised steel, joined with a special PVC adaptor. The upper part of the pushrod forms the 'walking stick' handle. The pushrod can be withdrawn through the cylinder from the surface. These features make dismantling and assembly very easy for a trained or technically-minded person.

The valves are made of rubber balls encased in PVC shells. These do not form perfect seals but are very reliable. The lower footvalve is screwed into the base of the cylinder and the piston valve attached to the pushrod. The difficult PVC/steel joints have been researched extensively and hold well despite the difficulty of matching steel and PVC. The steel handle is raised and lowered through a steel handle guide, fitted with a set of replaceable polyurethane bushes which should be changed twice a year. Longer lasting nylon bushes are planned for future use.

Characteristics

The Mark I range delivers 30–40 litres of water per minute and the Mark VI 15 to 20 litres per minute. The Mark I is suitable for pumping from wells down to a maximum of 6 metres depth and the Mark VI from wells down to 12 metres. The Mark I is a strong pump and can be used in heavy duty settings by over 100 persons. The Mark VI is designed for use either by a single family or small group of families (maximum 60 persons). Pump life varies considerably, depending on the care taken on assembly and installation and the number of users. Earlier models were designed with a threaded head assembly which often became loose when heavily used. In a communal setting, the pump should last for five years, but may extend well beyond ten years in a family setting. Polyurethane guides wear fairly rapidly (nylon would be a better material), and should be replaced every six months in extended family or community settings or yearly in family settings. If the polyurethane guides are not replaced, the steel handle will wear against the steel handle guide, leading to wear in both. Piston valves should be replaced when the delivery rate is reduced — probably every two or three years. However, in family settings they are known to operate for over five years without needing replacement. Apart from cleaning the pump, which should be done annually, replacing the polyurethane guides and ensuring that the head bolts are tight are the only routine maintenance procedures.

The Blair Pump on a protected well.

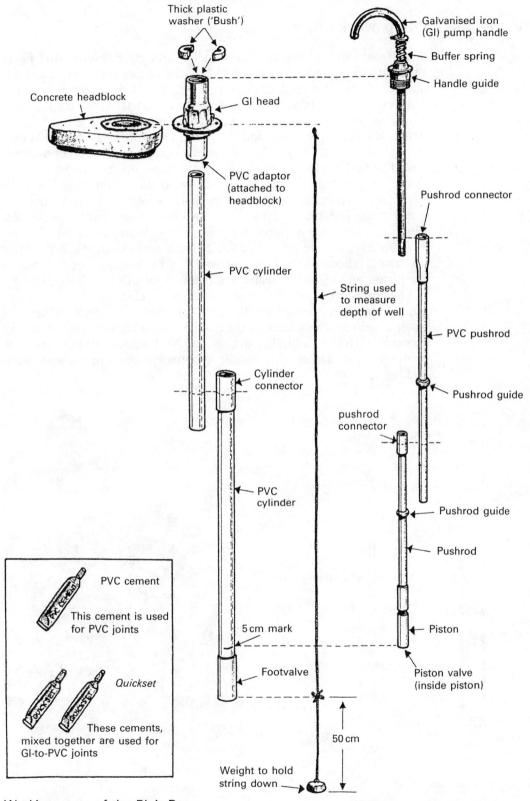

Working parts of the Blair Pump.

113

How the Blair Pump works

When the pump is installed in a well, water rises through the lower valve into the cylinder and also bypasses the piston valve. The handle is now raised and lowered to deliver water. When the handle, which is connected to the pushrod, is raised, water is drawn in through the open footvalve, whilst the piston valve remains closed. In the critical downstroke, the footvalve closes and the piston valve opens as it passes down through the cylinder. As the two valves come closer together water is forced to find a way out, the easiest path being up the hollow pushrod to the surface. Some water is forced around the piston, which has the advantage of lubricating the moving parts whilst they are in motion. For 6 metre lifts the pump works with worn valves and leaky pistons. Finer tolerances are required for greater lifts. Water is effectively used as a lubricant on the critical piston/cylinder surfaces in much the same way as oil lubricates the same surfaces of a car engine. For this reason the piston and cylinder wear away remarkably slowly.

The force used to pump water is directed close to the line of the pump, and these forces are buffered by the water itself. When the water is delivered to a bucket, the handle cannot come to the limit of its downward stroke, and this avoids the 'end knock' characteristic of many pumps. The pump handle is situated close to ground level, making it easier for small children to use.

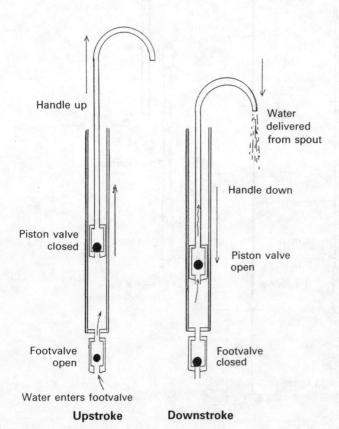

The Blair Pump lies close to the ground and is convenient for use by children.

Installation of the Blair Pump Mark VI

This is the most widely used Blair Pump, and is sold in lengths which range from 2 to 12 metres. In common with all handpumps, it can only work at its best when properly installed. Correct assembly of the parts is also vital. Certain parts of the pump will wear out in time, and it is important, when ordering pumps from the manufacturer (Prodorite Ltd, Harare) to order a few essential spare parts. Tools are provided with each pump. No pump will work for ever without maintenance and repair and the Blair Pump is no exception. Wherever the Blair Pump is installed provision must be made for maintenance and repair. The parts of the pump are illustrated in the diagram.

The new Blair Pump head block is designed to bolt on to a steel baseplate support cast in a concrete slab. Earlier Blair Pump head blocks with threaded inserts tended to wear in heavier duty settings, although they kept tight in lighter duty family and extended family settings.

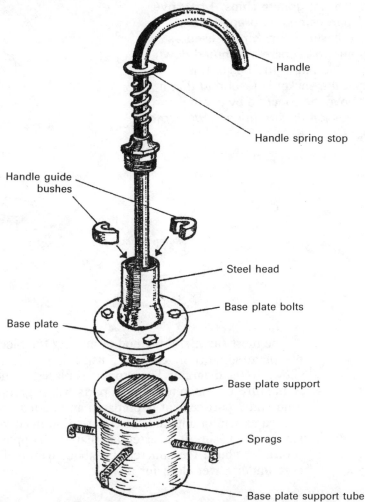

The Blair Pump can be fitted to either a wide diameter well or a tubewell drilled by hand. Preferably the well should be lined from the bottom with burnt bricks, stones or well liners, and the lining raised about 300 mm above ground level, covered with a strong concrete slab and surrounded by a well-made apron and water run-off channel. When fitted on a well, the well slab should be made with a central hole large enough to accept a bucket in case of emergencies. The concrete head block of the pump is mounted over this well slab and mortared in position.

Casting the concrete head block

The concrete head block is a very vital part of the pump and must be very strong. A high quality commercial head block is available and when used makes a neat and secure installation. Normally pumps are sold with a steel pump head baseplate support which is designed to be embedded in a strong concrete headblock.

The baseplate support tube is placed in the middle of the basin. A mixture of concrete using 2 parts 6 mm granite chips, 1 part river sand and 1 part cement is prepared and added to the basin. 3 mm wire is used as reinforcing and the concrete rammed down well and levelled flat in the basin. It is essential that the socket is level and that it protrudes above the concrete by about 6 mm. The concrete is left to cure under a wet cloth for a week.

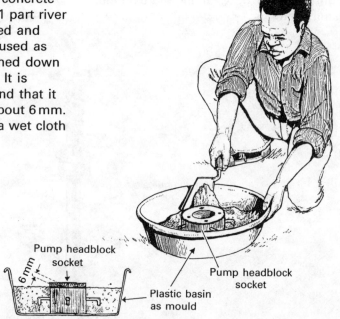

The most successful method of making the block involves casting the baseplate support in concrete, using a plastic basin as a mould, as shown in the diagram. The support is placed in the middle of the basin. A mixture of concrete using 2 parts 6 mm granite chips, 1 part river sand and 1 part cement is prepared and added to the basin. 3 mm wire is used as reinforcing and the concrete rammed well down and levelled flat in the basin. It is essential that the support is level and that it protrudes above the concrete by about 6 mm. The concrete is left to cure under a wet cloth for a week.

Mounting the concrete headblock

The headblock is removed from the mould, washed and cleaned and mounted in cement mortar over the well or tubewell, as shown in the diagrams. It is very important to ensure that the head block is level and securely concreted in place, and must lie directly above the opening in the well slab. Follow these steps so that the headblock is correctly positioned.

1
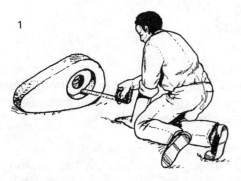

Measure the depth of the pump socket hole in the concrete headblock (approx 10 cm/4 inches).

2

Record this measurement on the PVC casing.

3

Cut the PVC casing to size.

4

Lay cement mortar around the PVC casing. Use a spirit level to position the headblock correctly. The PVC casing must fit centrally inside the headblock.

Assembly of the pump

It is essential that the pump is assembled carefully, and that the inner and outer pipes are cut to match each other. First the well or tubewell is measured for depth with a weighted string. The depth should be measured from the top of the concrete head block.

Concrete headblock

Correct measurement of the well or tubewell is very important. The pump is made 500 mm shorter than the well depth.

Next the string is laid on the ground next to the well and the pump sections are laid alongside the string. The cylinder pipe is cut and the sections cemented together so that the footvalve is 0.5 metres from the bottom of the well. It is very important to use fresh PVC cement, and this can be provided in cans or in small tubes. The upper length of cylinder is cemented into the PVC adaptor attached to the pump head. The cylinder pipe is cut and cemented together with PVC sockets (connectors) to the correct length. A hacksaw can easily cut the pipe. Once the cylinder has been finished, the pushrod is assembled in the same way as follows:

The important joint between the steel pump handle and the PVC pushrod is screwed together so that the full length of the thread is used. On many pumps a locknut is provided to secure this section. It is better, however, to cement this joint with a fast-setting epoxy resin cement (Trinepon 6). Once cemented together it is very unlikely that this joint will ever separate. The correct length of the pushrod is determined by placing the steel handle guide alongside the pump head as shown in the diagram, and with the handle in the down position the lengths are assembled, cut and cemented together so that the lower end of the piston valve is opposite the 5 cm mark on the outside cylinder pipe. It is very important that these two pipes are cut to the correct length. Once assembled the valves are removed, inspected, washed and re-assembled and the whole pump thoroughly washed down with water.

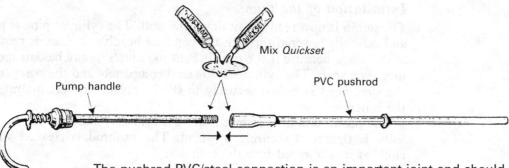

Mix Quickset

Pump handle

PVC pushrod

The pushrod PVC/steel connection is an important joint and should be epoxy cemented together.

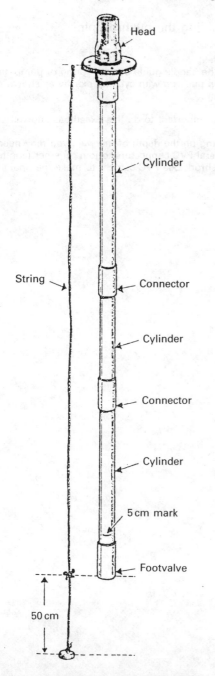

Assembling the cylinder pipe.

Join galvanised iron headblock to PVC cylinder.

* Use PVC Cement for this task.
* Clean joint ends thoroughly.
* Clean away extra glue.
* Leave to dry.

Join PVC cylinder at footvalve end.

Make final cylinder join using a connector.

To do this:
* Measure the cylinder length exactly as shown in the illustration.

When completed the PVC cylinder is 50 cm shorter than the depth of the well.

The number of connectors needed depends on the depth of the well.

Installation of the pump

The pump is now ready to fit down the well. The cylinder pipe is taken and carefully lowered through the concrete headblock. Care is required at this stage because if the pipe is bent too much it may be damaged or may even snap. The cylinder is lowered completely and the baseplate of the steel head is bolted securely to the baseplate support mounted in the concrete.

The pushrod is now inserted into the cylinder, making sure that the valve is tightened securely by hand. The pushrod is cleaned with a cloth as it is lowered into the cylinder.

Assembling the pushrod pipe.

To do this:
* Push the handle guide against spring of pump handle.
* Line up pushrod with cylinder exactly as shown in the diagram.

It is very important to do this exactly as shown.

Depending on the depth of the well, you may need to use several PVC connectors to join together lengths of PVC pushrod. Use PVC cement to make the final joins.

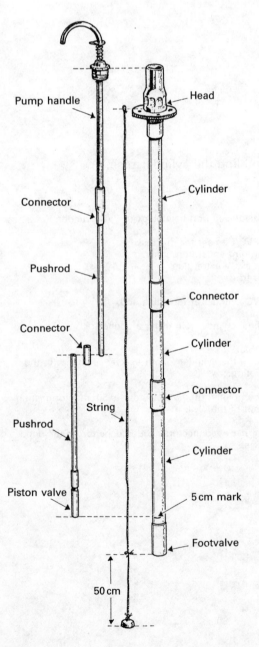

Before the handle guide is screwed into the steel head, two polyurethane bushes are inserted into the head as shown in the diagram. These bushes prevent metal to metal contact at this point. Finally the handle guide is screwed into the head tightly with the spanner provided.

The pump can now be tested. Water should be delivered from the spout after three or four strokes have been made with the handle. A long stroke with pressure being exerted on the downstroke is best. Ideally both hands should be used for pumping. The correct method of using the Blair Pump is soon developed by the users.

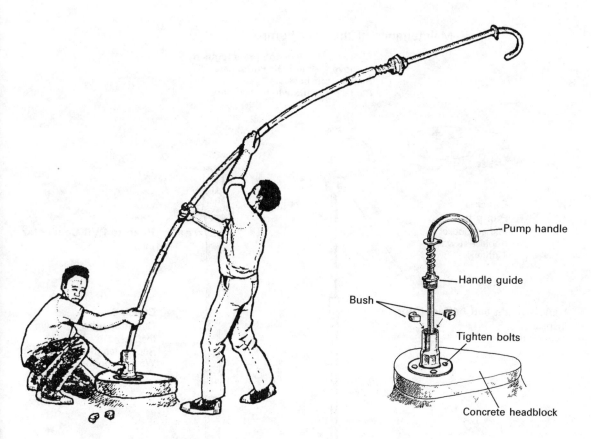

Carefully lower the pushrod into the cylinder.

Maintenance of the pump

Every Blair Pump should be served by an effective maintenance system. Each pump can only be expected to give long reliable service if spare parts are available, and people equipped with tools are trained in maintenance and repair.

It is not unusual for a minor problem to develop with any pump. A loose thread or blocked valve can result in complete failure. Very often these faults can be repaired with great ease, especially with a simple pump like the Blair Pump, but without training and knowledge, even the simplest fault can lead to a pump being removed and rejected.

Thus in any programme of providing water with the Blair Pump, or indeed any pump, it is essential to:

1. Provide the necessary tools.
2. Provide the necessary spares.
3. Provide the necessary training.

Each pump is provided with a spanner, and a maintenance and repair kit is also available.

Maintenance of the Blair Pump.

Check.....all working parts regularly
Remove....the Blair Pump carefully
Repair....if possible
Replace...parts when necessary

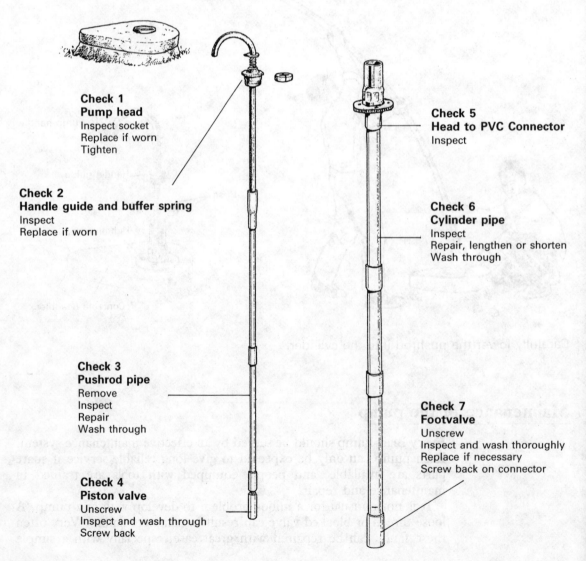

Check 1
Pump head
Inspect socket
Replace if worn
Tighten

Check 2
Handle guide and buffer spring
Inspect
Replace if worn

Check 3
Pushrod pipe
Remove
Inspect
Repair
Wash through

Check 4
Piston valve
Unscrew
Inspect and wash through
Screw back

Check 5
Head to PVC Connector
Inspect

Check 6
Cylinder pipe
Inspect
Repair, lengthen or shorten
Wash through

Check 7
Footvalve
Unscrew
Inspect and wash thoroughly
Replace if necessary
Screw back on connector

Parts requiring maintenance

The following parts require routine inspection for looseness and wear. It is important that all loose parts are tightened and worn parts replaced at the earliest possible time. Failure to do this may result in premature replacement of parts, which leads to unnecessary expense.

1. Steel head

This is a critical part of the pump and the bolts used to secure the head must remain tight. On older Blair Pumps it is essential to ensure that the threaded head assembly is kept tight, as these parts are not cheap to replace. It therefore makes sense to keep the parts tight.

2. Handle guide bushes

These are made of polyurethane and are designed to take up wear in the handle guide. If the handle wears against the steel handle guide, both the guide and the handle wear away. If this wear is allowed to continue, the opening in the steel guide becomes greatly enlarged and the wall of the handle becomes thinner. If this is allowed to continue both parts will require total replacement. The polyurethane bushes are designed to avoid this problem. However they do require replacement at least once a year.

3. Steel handle guide

This part should be kept tight to retain the polyurethane bushes in place.

4. The pushrod and piston valve

Under recommended working conditions these should last for several years before requiring replacement. However, as the piston valve wears thinner, the efficiency of the pump is reduced. Also debris from the well can partially block a valve and reduce the output or stop it altogether. It is a simple matter to unscrew the handle guide and completely remove the pushrod and piston valve to check on wear and to clean the whole assembly. Deposits do build up on the pushrod and these should be cleaned off periodically. When the piston valve is worn so that it seriously reduces the output, it should be replaced with a new unit.

5. The cylinder and footvalve

The cylinder should not give trouble if it has been well cemented together, but some care must be taken when it is inserted and removed, especially when it is older. Without care, the pipe may crack and require joining with a PVC socket. If the pump is installed on an older well, it is possible that bits of rope or other debris which are left in the well may be drawn up into the footvalve and seriously impair its performance. Very often faulty pumping is caused by pieces of debris blocking the valve. If this is a serious problem, the well should be cleaned out or a piece of stainless steal gauze fitted over the end of the footvalve. As with the pushrod, the cylinder should be thoroughly cleaned and the valve tightened before installation.

Repair

Those who are trained in Blair Pump repair should be able to attend to the following potential points of breakage or wear.

1. Broken pushrod (shortening and lengthening)
2. Broken cylinder (shortening and lengthening)
3. Damaged valves and replacement
4. Damaged footvalve connector and replacement
5. Damaged cylinder adaptor and replacement
6. Loose pump head base plate support
7. Worn steel head and replacement
8. Worn handle guide and replacement
9. Worn handle and replacement
10. Worn galvanised iron to PVC pushrod adaptor

These will be dealt with in more detail.

Blair Pump Mark VI. Spare parts.

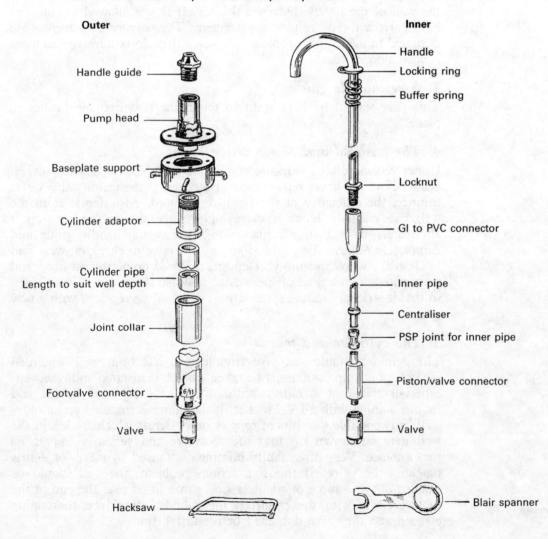

1. & 2. Repair or length change of pushrod or cylinder

The entire pump should be removed, cleaned and the broken section of the pushrod or cylinder sawn off straight on both broken sides. A 25 mm socket for the pushrod should be taken (or 50 mm for the cylinder) and the joint reconnected using PVC cement.

To repair or change length of PVC pushrod or cylinder:

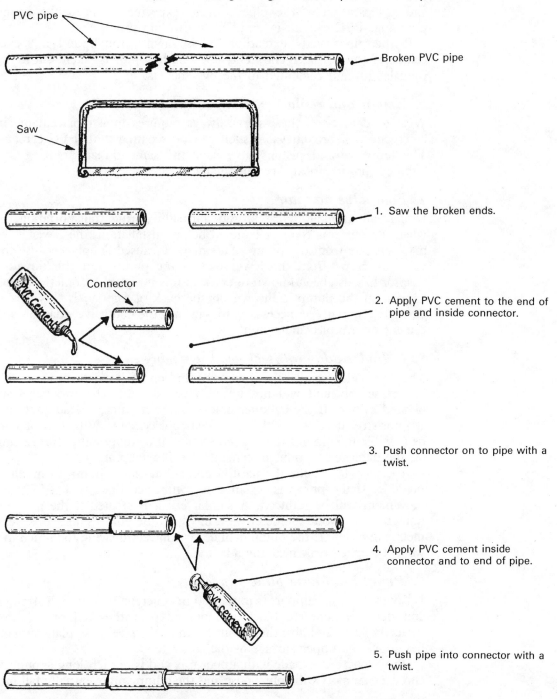

1. Saw the broken ends.
2. Apply PVC cement to the end of pipe and inside connector.
3. Push connector on to pipe with a twist.
4. Apply PVC cement inside connector and to end of pipe.
5. Push pipe into connector with a twist.

Use of PVC cement

If possible the pipe and the socket should be sandpapered on the surfaces which will come into contact. Fresh PVC cement is then applied to both surfaces so that they are well covered. The pipe is then pushed into the socket as far as it will go, and given a half turn. The joint should be given at least 10 minutes to set.

It is essential that fresh cement is used. Old cement stored in a can, and which has become stiff will not bond the pipes together properly and a separation of the pipes can be expected — probably during pumping. PVC cement is supplied in tins and tubes.

If the pipe needs extending, it is sawn through and a socket cemented to each sawn end. The new length of pipe is taken, cut to the right length and cemented to the two sockets.

3. Repair and replacement of valves

Wear or damage to these parts always requires total replacement. In emergencies a broken valve shell can be repaired with PVC cement. The footvalve and piston valve have the same threaded fitting and although not identical, are interchangeable.

4. Footvalve adaptor

In a small number of cases, and normally during transportation, or when the pump is roughly handled before installation, the lower valve may separate from the pump. This may be caused simply because the valve unscrews from the lower end of the pump, but the threaded adaptor has also been known to crack in this position — something that happens if the pump is thrown off the back of a lorry. If this part is damaged, it will be necessary to saw off the footvalve adaptor and cement on a replacement unit.

5. Cylinder adaptor/Steel head assembly

The PVC cylinder adaptor is screwed and epoxy-bonded into the steel head assembly and will not unscrew or fall out. The two parts are bonded as one. If the cylinder cracks where it enters the adaptor, the adaptor/steel head assembly may need replacement. With care it may be possible to chip out the broken section of cylinder and, after sawing off the damaged section, re-cement it to the adaptor.

It is wise to reduce the natural oscillation of the pump as much as possible. If the pump is installed in a tubewell lined with PVC, this movement will be reduced. A similar effect is created if the pump is immersed in a wide diameter well, in which the pump stands in several metres of water. If the pump cylinder is allowed to swing too much in a well, it might break near the adaptor.

6. Pump head base plate support

It is vital that the support is immersed in concrete of very high strength and that the concrete block is mounted over the well or tubewell perfectly level, and also that about 6 mm of the steel base plate support lies above the upper surface of the concrete.

If the concrete is made with a poor mix and is not left long enough to cure, the support will come loose and the pump will eventually fall into the well. Properly-made concrete blocks should be cured under water for at least one week, and should be secured on the well head in strong cement mortar, which should be left to cure for several days before the pump is used.

7. Steel head assembly

It is essential that the steel head is kept tightly bolted to the baseplate support in the concrete head block.

Tightening the head.

8. Handle guide

If the hole in the guide through which the handle passes becomes too large, the guide must be replaced. This can be avoided by the regular use of the polyurethane bushes which are enclosed in the guide. Spare sets of bushes should be made available with each pump.

Tightening the handle guide.

9. Handle

With care the handle should last for several years. The handle can be bent if it is abused, however. With normal wear, the wall thickness of the handle decreases as the years pass by. The wall thickness in a heavily used pump may be reduced to the point where the handle must be replaced every 3–5 years. The use of the polyurethane bush reduces wear on the handle.

10. Pushrod iron to PVC adaptor

This appears to be a robust unit, and lasts well if the joint is cemented with fast-setting epoxy resin. If a crack appears or the joint gives way, total replacement is essential. A certain amount of wear will occur along the shaft of this adaptor, but several years of life can be expected.

Inserting the pushrod and handle.

Excercises in Blair Pump maintenance

A number of practical exercises can be carried out to improve the skill of those who are responsible for Blair Pump maintenance and repair. A small number of trainees are equipped with a repair kit and asked to repair a pump which has been deliberately broken or in which the working parts have been jammed or blocked deliberately. A trainee has then to repair the pump under the watchful eye of the trainer and the other trainees. Their knowledge is then tested. It is during such exercises that trainees learn fastest.

Exercises

1. Joining 25 mm and 50 mm PVC pipe

This technique is important and requires PVC pipes, suitable PVC sockets and PVC cement. If sockets are in short supply and cannot be spared the PVC cement can be substituted by a mixture of flour and water. However it is best for all trainees to cement joints as they would in the field.

2. Dismantling and reassembling the pump

All trainees should be able to unscrew the handle guide and remove the pushrod pipe and then separate the handle, GI to PVC adaptor and valve. They should also be able to remove the cylinder pipe and remove the footvalve. All parts should then be reassembled and fitted back into the well.

3. Blocked valves

Trainers can fill the valve chamber with bits of rope, so the performance of the pump is poor. Using his maintenance kit, the trainee is asked to repair the pump and restore original efficiency.

4. Broken pipes

A working pump can be removed by the trainer, and the inner and outer pipes partly cut in two. The broken pieces can then be replaced down the well. The damaged pump is then tested and withdrawn by the trainee who:

a) Removes the parts from the well.
b) Checks for the damaged parts and wear.
c) Checks that the pushrod and cylinder are the correct lengths and adjusts if necessary.
d) Rejoins the broken pipes.
e) Refits the pump.

5. Pump too short

Pumps that do not reach the water cannot pump. This is an exercise in pump lengthening.

6. Pump too long

The pump footvalve is embedded in the sand at the base of a well. This can be prepared by the trainer. The trainee extracts the pump, assesses the problem, measures the depth of the well, adjusts the length of the pump and cleans and refits the pump.

Extending the cylinder.

7. Piston valve worn out

The rate of pumping is reduced by the trainer, by replacing the normal piston valve with one that has been removed from an old pump. The trainee removes and examines the valve and makes the necessary correction — in this case replacing the valve.

8. Setting the base plate support in concrete

If possible a pump with a defective or worn-out steel head is placed in the training exercise, the trainee is asked what is wrong and then asked to embed a new support in concrete within a suitable mould.

Casting the head block.

9. Replacement of old steel head assembly

The pump is removed and the old cylinder pipe cemented to a new cylinder adaptor in a new steel head. A new base plate support is mounted in a new concrete head block as in 8. The pushrod is adjusted in length and after the concrete parts are cured the pump is refitted.

10. Sand in the pump

A Blair Pump cannot operate if sand has been introduced into the cylinder. Sometimes in transit, a pump gets sand or grit into the working parts, and in some cases the well or tubewell has sediment in it which pass into the pump and affect the performance. This exercise is designed to emphasise pump cleanliness. The pump is removed, thoroughly washed down and cleaned and dismantled. It is then refitted together and placed down the well.

Installing a Blair Pump into a hand-drilled tubewell

Two hundred and fifty Blair Pumps were fitted into the peri-urban settlement of Epworth, close to Harare, in 1985/86 within a few months using the following technique, which can be used as a training excercise.

Stages in operation

1. Locate site with local villagers, health workers and by water divining.
2. Assemble drilling rig at the site and hand-drill tubewell down to bedrock — at least 3 metres of water are required.
3. Prepare 3 metre lengths of 110 mm class 6 PVC casing, with the lowest length slotted. A 1.5 metre slotted section may be adequate.
4. Cement on PVC end cap to the end of the slotted pipe and then lower casings down one by one, cementing pipes together, until the full depth of the drilling has been cased.
5. Cut off the casing 300 mm above ground level.

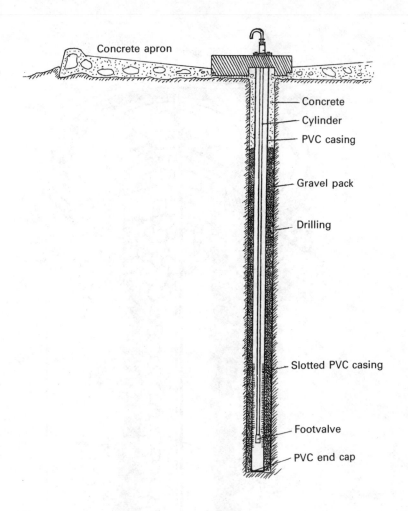

6. Add gravel pack of 6 mm granite chips or very course river sand (over 2 mm) to the annular space between drilling and casing to within 1 metre of surface.
7. Fill remaining metre of annular space with concrete
8. Lay a wide circle of bricks or rocks 3 metres in diameter to form the rim of the apron around the tubewell. Also lay bricks along the course of the 6–10 metre water run-off channel down hill from the apron.
9. Add and shape the concrete to form the apron and water run-off so that all waste water drains into the seepage area at the end of the channel.
10. Prepare the concrete head block for the pump and when cured fit in cement mortar over the tubewell.
11. Assemble the pump to suit the depth of tubewell, so that the footvalve is 500 mm from the bottom of the well.
12. Inspect and clean the pump and fit it into the tubewell.
13. Villagers should be shown all stages and encouraged to participate.
14. A spanner should be left on site to enable the chosen pump caretaker to keep the steel head tight.

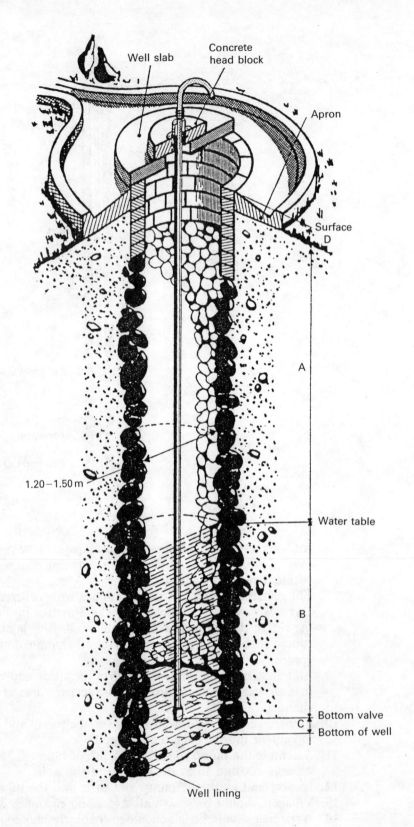

Blair Pump placed in a well.

Handmade Blair Pumps

Although most Blair Pumps used in Zimbabwe are commercially made, the handyman can fabricate his own pump, which can provide good service, especially in the homestead. Standard PVC pipes and plumbing fittings can be used for this job. Two models of Blair Pump have been designed for handmade construction; a lighter duty family model and a heavier duty model, which is useful for families or communities. Both operate down to a depth of 6 metres. This pump can be made in the school classroom, as it does appear in the school curriculum and 'O' level syllabus.

The light duty Blair Pump

This pump was designed by Blair Research Laboratory in 1983 for light duty family use. It is made from PVC and galvanised iron fittings, with more PVC and less iron than in other Blair Pumps. In common with all Blair Pumps, the handle doubles as a pushrod and water spout — water comes out of the handle on the downstroke.

Basically the pump consists of two PVC pipes, one running up and down inside the other. Each pipe is fitted with a simple non-return valve at its lower end. The lower end of the inner pushrod is expanded to form a piston with a non-return valve fitted inside it. The upper end of the pushrod forms a handle and spout. The piston forms a loose fit within the outer pipe which acts as a cylinder. The outer cylinder pipe is hung from the top of the well from a concrete head block. In the light duty model the footvalve can be extracted through the cylinder pipe.

When the pump is installed into a well, water rises first through the lower footvalve and then through the upper piston valve. When the handle is raised water is drawn in from the well into the pump cylinder through the footvalve. When the handle is pushed down, the lower footvalve closes and the upper piston valve opens. The water between the two valves is now forced to find a way out as the piston is lowered towards the footvalve. It tends to rise up through the piston valve and up the hollow pushrod to the surface and out through the handle. Some water bypasses the piston, but most passes up the pushrod. In this way water is pumped from the well to the surface.

How to make the pump

A number of standard steel and PVC pipes or fittings will be required. These are:

Steel fittings
1. 1 × 40 mm galvanised iron socket
2. 1 × 40 mm to 25 mm (1½" to 1") galvanised iron reducing bush

PVC fittings/Pipes
1. Suitable length of 40 mm class 10 PVC pipe (cylinder)
2. Suitable length of 25 mm class 26 PVC pipe (pushrod)
3. 1 × 25 mm PVC tee

4. 1 × 40 mm PVC thread to pipe adaptor
5. 4 × 25 mm thread to pipe adaptors (internal 20 mm thread)
6. 3 × 25 mm thread to pipe adaptors (external 20 mm thread)
7. 1 × 32 mm thread to pipe adaptor (external 25 mm thread)
8. 2 small glass marbles
9. 2 brass washers
10. Small disc of stainless steel screen (optional)

Parts of the pump

These include:

1. The pump head
2. The cylinder (including lower valve anchor)
3. The pushrod (including handle and pushrod guide)
4. The piston valve
5. The footvalve
6. The lower valve extractor

Parts of the family Blair Pump.

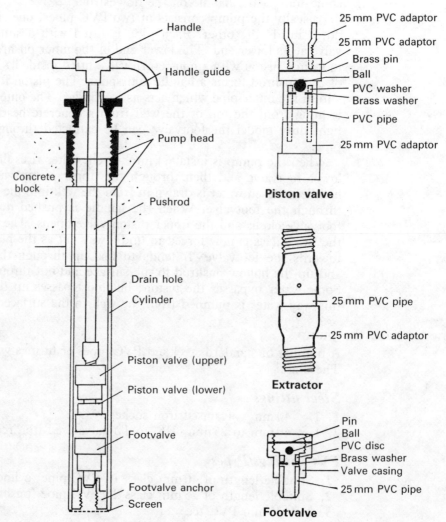

1. The pump head

This consists of a 40 mm PVC thread to pipe adaptor (with external thread) tightly threaded into a steel 40 mm socket. These two units are cast within a strong block of reinforced concrete. One method of doing this is to place the steel and PVC assembly into the middle of a plastic bowl (steel part upwards) and fill it up with a strong concrete mixture (2 parts gravel, 1 part river sand, 1 part cement) reinforced with 3 mm wire. The mixture is levelled flat in the bowl so that approximately 6 mm of steel socket protrudes above the concrete level. The concrete is allowed to cure for a week and kept wet.

2. The cylinder

This consists of a suitable length of 40 mm class 10 PVC pipe, about 500 mm less than the depth of the well. A lower valve anchor is cemented with PVC cement into the lower end of the cylinder pipe. The anchor consists of a 25 mm PVC pipe to thread adaptor (internal thread), which is lathed or ground down so that it will fit inside the bottom of the cylinder with the threaded end facing downwards. The adaptor is cemented in position. The lower valve will press fit into the other end of the adaptor.

3. The pushrod, handle and pushrod guide

The pushrod consists of a length of 25 mm class 26 PVC pipe, approximately the same length as the cylinder. This will be recut later to the exact length. A handle is made on one end of the pipe. First cement a 25 mm PVC tee on to the end of the pipe. Next cut two short lengths (80 mm and 150 mm) of 25 mm class 26 pipe. Insert a plug of hard-setting putty (Trinepon 13) into the shorter length and cement this into one of the arms of the tee. The other length of 25 mm pipe should be bent as shown in the diagram. The pipe should be heated over a flame till it becomes soft and then carefully bent through about 45 degrees. Some practice may be required to do this. Once bent, this piece of pipe is cemented into the other arm of the tee with the outlet facing downwards.

The pushrod guide is made by taking the 40 mm to 25 mm galvanised iron reducing bush and tightly threading into it the 32 mm PVC thread to pipe adaptor (this has a 25 mm external thread). The PVC adaptor is sawn off approximately 6 mm above the threaded section. The 25 mm PVC pushrod should slide neatly through this PVC guide.

4. The piston valve

This is made up of two 25 mm PVC thread to pipe adaptors (internal thread), some short lengths of 25 mm class 26 PVC pipe, a brass washer and a small glass marble. When complete the whole assembly screws into a 25 mm PVC adaptor (external thread) which is cemented to the lower end of the pushrod.

Both the 25 mm PVC adaptors must be lathed or ground down so that they will fit within the 40 mm class 10 cylinder pipe, and will be able to slide easily. Next a PVC washer 8 mm deep is cut exactly straight from the 25 mm class 26 pipe. This is cemented within the pipe end of the 25 mm adaptor and pushed up as far as it will go. Next the brass washer (outer diameter 23 mm, inner diameter 12 mm) is inserted within the adaptor so that it sits on the PVC washer as shown in the diagram. Next a length of 25 mm class 26 pipe (45 mm long) is cemented behind the brass washer so as to sandwich it between PVC. The brass washer is centralised before the PVC cement dries. The second 25 mm adaptor is cemented to the 25 mm class 26 pipe as shown in the diagram.

The next stage involves drilling two 3 mm diameter holes in the upper 25 mm adaptor through which a retaining pin will be pushed. The exact positions of these holes must be made by fitting the marble into the valve so that it sits on the brass washer. The marble must be given sufficient movement to allow water to pass it, yet be held by the pin so that it is not forced out of position. After the holes have been drilled the marble is placed in position and checked to see if it fits up against the brass washer which acts as a valve seat. A length of 3 mm brass or steel rod is inserted as a retaining pin and cut off flush with the adaptor. Finally a 25 mm PVC adaptor with external thread is screwed into the upper part of the valve. This adaptor will be cemented to the pushrod at a later stage.

5. The footvalve

This valve is made in exactly the same way as the piston valve, but the lower 25 mm PVC adaptor (internal thread) is not fitted. The length of the 25 mm class 26 pipe which forms part of the lower end of the valve is bevelled so that when it is pushed down the cylinder it will make a neat push fit into the lower valve anchor.

6. The lower valve extractor

This consists of two 25 mm PVC adaptors (external thread) which are cemented together as shown in the diagram with a short length of 25 mm class 26 pipe.

7. Assembly of the parts

The pump should reach down to within 500 mm of the bottom of the well or tubewell and the length of the cylinder is cut accordingly and cemented into the PVC socket in the lower side of the concrete head block.

The pushrod (with pushrod guide in place) is laid alongside the cylinder and the piston and footvalve are placed in their correct positions so that the correct length of the pushrod can be judged. When fitted, the two valves should be about 50 mm apart. Once the length of the pushrod has been checked, it is cut off to the right length and cemented to the piston valve through the adaptor.

The whole pump is washed thoroughly with water and then the lower valve can be fitted. To do this, the lower valve extractor is screwed into the lower end of the piston valve tightly and the footvalve is screwed on to the extractor loosely. All the valves are now pushed down the cylinder and the footvalve pressed into the lower valve anchor — this should be a neat push fit. The pump handle is now turned anti-clockwise and this should unscrew the footvalve from the extractor. Next, the pushrod is removed and the extractor taken off the piston valve. The pushrod is placed back within the cylinder pipe and the steel bush screwed into the steel socket in the head block. The pump is now ready to insert in the well.

Installation

A concrete well cover should have been prepared to fit over the well and this should be cemented in position. The well cover should have a hole left in it approximately 100 mm across through which the pump can be inserted. The head block with pump attached is carefully raised up and the pump lowered through the hole in the well cover. The head block is cement mortared in position on the well head and left to set. An alternative technique is to cement the cylinder into the head block once the pump has been placed down the well. The pump is now ready for use. The lower valve can be inspected by removing the pushrod and attaching the extractor to the piston valve and threading this on to the footvalve by turning the handle in a clockwise direction and pulling.

Maintenance

If used by a small family the pumps should last a few years. Obviously if parts wear they should be replaced. The pump is quite cheap to make and works well on shallow wells. It is a good model to make in a classroom and demonstrates how water can be raised in a pump from a family well.

The heavy duty Blair Pump

This was designed by the Blair Research Laboratory in 1976 for use in community settings on shallow wells. It is a heavy duty pump, but has a maximum lift of 6 metres. Several models were placed on trial between 1976 and 1980. The following account describes the construction of the later Mark Ic model.

How to make the pump

A number of steel and PVC fittings and pipes are required as follows:

Steel/Brass fittings

1. 1 × 25 mm galvanised iron (GI) bend
2. 1 × 25 mm GI tee
3. 2 × 25 mm GI barrel nipple
4. 1 × 75 mm GI socket
5. 1 × 75 mm to 50 mm GI reducing bush
6. 1 × 50 mm to 32 mm GI reducing bush
7. 1 metre length 25 mm GI pipe
8. Up to 5 metres 20 mm GI pipe
9. 1 × 25 mm to 20 mm reducing socket
10. 20 mm brass flap valve
11. 1 × 50 mm GI barrel nipple
12. 1 × 50 mm GI socket
13. 1 × 50 mm footvalve
14. 1 × 100 mm nut and bolt

PVC fittings/Pipes
1. Suitable length 90 mm class 16 PVC pipe (cylinder)
2. 1 × 90 mm PVC thread to pipe adaptor (75 mm external thread)
3. 2 × 75 mm lengths of 75 mm class 10 PVC pipe
4. 1 × packet hard setting putty (Trinepon 13)

Parts of the pump
These include:

1. The pump head (including pushrod guide)
2. Pump cylinder
3. Pushrod assembly including handle
4. Piston valve
5. Footvalve

The heavy duty handmade Blair Pump.

The heavy duty Blair Pump is designed for handmade production. It is suitable for shallow wells down to a maximum depth of 6 metres.

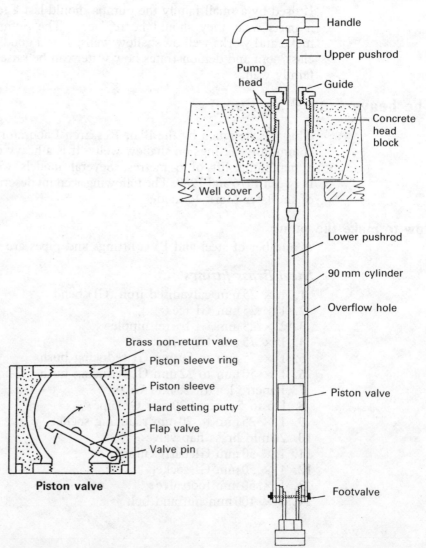

1. The pump head

This consists of a 90 mm PVC thread to pipe adaptor tightly threaded into a 75 mm galvanised iron socket. These two units are placed in the centre of a high plastic bowl and a strong mixture of wire reinforced concrete (2 parts gravel, 1 part river sand and 1 part cement) is added. The socket is placed upper-most and the concrete added and levelled to leave 6 mm of the socket protruding above the concrete level. This is left to cure for one week. The pushrod guide consists of a 75 mm to 50 mm GI reducing bush into which is screwed a 50 mm to 32 mm reducing bush. The smaller bush is lathed down so that the 25 mm pushrod will freely pass through it. The double bush is used because 50 mm steel fittings are more readily available and cheaper than 75 mm fittings. The inner bush will eventually wear oval and should be replaced.

2. The cylinder

This consists of a length of 90 mm class 16 PVC piping which is a very tough material. There is some variation in the internal bore and a pipe must be chosen which a 75 mm PVC pipe will pass through. The piston valve assembly is confined within a 75 mm PVC sleeve.

The brass footvalve is secured with a lower valve anchor. The anchor is made by taking a standard 50 mm steel socket and pushing it into a 75 mm length of 75 mm class 10 PVC piping in a vice. This is made easier by tapering the socket. The anchor is then PVC cemented into the end of the cylinder, and held in position with a retaining bolt, which is drilled through the assembly as shown on the diagram. The footvalve is connected to the 50 mm socket through a barrel nipple.

3. The pushrod assembly and handle

This consists of standard 25 mm and 20 mm steel fittings. The handle consists of a 25 mm steel tee, two 25 mm barrel nipples (one of which is sealed off with hard-setting putty), and a 25 mm steel bend. Galvanised fittings are preferred. The upper pushrod is made of a one metre length of 25 mm steel pipe. The upper 25 mm pushrod is connected to the lower 20 mm pushrod with a 20 to 25 mm GI reducing socket. The lower pushrod is cut to the required length (see later). A thick rubber disc is cut and placed tightly around the upper pushrod (under the tee). This acts as a rubber stop.

4. Piston valve

Although the 20 mm (3/4") brass non-return valve will wear out in time, it is known to give good service and is easy to incorporate in the piston assembly. The length of the piston will depend on the length of the brass valve chosen—there is some variation, but it will normally be about 75 mm. First cut a 75 mm length of 75 mm class 10 PVC pipe (sleeve) so that the ends are cut straight. Next cut two rings of the 75 mm pipe 5 mm wide. Cut them so that they can be cemented inside each end of the 75 mm sleeve. These rings help to retain the putty which binds the brass valve to the sleeve. Before the brass valve is fitted within the 75 mm sleeve it must be trimmed down. This often necessitates sawing off the lateral inspection cap and grinding down the valve so that it will lie centrally within the PVC sleeve. Once the valve has been trimmed it is carefully fitted centrally within the PVC sleeve and a packing of hard-setting putty (Trinepon 13) is pushed in to fill the space between the valve and the sleeve. It is very important to perform this stage of the pump assembly with care and ensure the non-return valve lies exactly centrally. Once the piston valve has set it is ready for use.

5. *The footvalve*

A high quality brass footvalve is recommended for use, and is attached to the footvalve anchor at the end of the cylinder through a 50 mm GI barrel nipple. High quality footvalves of this type are capable of giving many years service without inspection. Many have been in use for decades without needing repair. In the event of the footvalve needing inspection, the pump cylinder must be raised from the well head.

Assembly of the parts

Once the depth of the well has been measured it is possible to assemble the parts of the pump to the required depth, which is normally 500 mm short of the well depth. The cylinder is cut to the correct length and laid beneath the head block on the ground. The footvalve is fitted to the lower valve anchor through a barrel nipple. The upper pushrod is fitted through the pump guide, and laid against the cylinder. The remaining parts of the pump are laid out on the ground. The upper pushrod is one metre long, the lower pushrod is now cut and threaded so that when the piston valve is screwed into position it will clear the footvalve by about 50 mm. The correct lengths of piping are important. Once cut to the correct length all steel fittings are tightly screwed up. The pushrod is now fitted within the cylinder and the guide tightened to the pump head. The piston should slide freely up and down the cylinder.

Installation

The pump must now be placed down the well and the concrete head block cement mortared in position. If the pump is short, the cylinder can be PVC cemented within the head block and the whole pump lowered as one unit into the well. Longer pumps may require that the cylinder is lowered first and cemented to the head block assembly after it is lowered. Once the cylinder has been mounted, the pushrod can be lowered and the handle guide tightened. The head block should be allowed to set in position for two days before the pump is used.

As with all pump installations it is wise to make a strong wide apron and water run-off channel at the water point.

Maintenance

This pump is quite robust, but parts will eventually wear out. These include the piston valve, the upper pushrod and the pushrod guide. All these parts should be available as spares. The valve becomes thinner as it is used more, and efficiency is lost, but this may take years. The upper pushrod will wear thin in time and also the hole in the guide will become enlarged. The piston valve, pushrod guide and the upper pushrod are not expensive parts.

The high quality footvalve may not require checking for several years — hence the advantage in fitting a good quality product right from the start. The 50 mm footvalve sold by Radiator and Tinning (Bulawayo) is well proven.

The Nsimbi Pump

This pump was introduced into Zimbabwe by the Lutheran World Federation in 1983, and originates in Malawi, where it is known as the Malawi Mark V Pump. It is similar to several other direct action pumps, which include the Tara (Bangladesh), PEK (Canada), Wavin (Holland) and Nira AF85 (Finland). The Nsimbi Pump has never been commercially built in Zimbabwe and is hand-made in the LWF workshops, Bulawayo.

Description

The Nsimbi Pump uses a lift pump action similar to the Bush Pump. It is a direct action pump with a high content of PVC. A novel feature is the use of hollow pushrods, which become buoyant and ease the burden of pumping. The footvalve is a standard brass type, and the piston/plunger is a simple disc valve fitted to an imported double acting rubber seal. The pump head is made of cast steel, and is bolted to the well head. Originally the 20 mm tee-shaped handle passed through a nylon bush threaded into the pump head but this was replaced with a telescopic guide system. However, the former system proved to be more reliable and was re-established. The pump cylinder is made of 50 mm class 16 PVC pipe, the rising main of 50 mm class 10 pipe and the pushrods of 32 mm class 16 pipe. The water outlet is static and made of 40 mm steel piping.

The piston seal can be removed through the rising main for inspection and replacement. The pump head screws into the baseplate with a rubber gasket clamped between the two. A specialised steel to PVC adaptor has been designed for the crucial joints in the pushrod.

Nsimbi Pump.

Characteristics

The Nsimbi Pump delivers between 15 and 30 litres of water per minute depending on pumping head and rate of pumping. The pump can be fitted on to wells down to 13 metres in depth. The pumping action is easy when the seals are new and the rods buoyant. When the seals are worn, the delivery rate is reduced. Pumping becomes harder if the hollow pushrods fill with water due to leaky seals.

When placed in a light duty communal setting, pumps of this type have a life of about five to six years. The seals may last for two to three years. As with many direct action pumps, it is important to ensure that the pump head is kept tight, as this has a tendency to loosen, the pump head of the Nsimbi Pump is made of cast steel, and the threading comes under a great deal of strain if the water outlet pipe is not supported. If the outlet pipe is supported, the pump head itself becomes much more stable.

How the Nsimbi Pump works

This operates on the same principle as all lift pumps. Water is raised through the rising main as the piston valve is raised through the cylinder. The piston is a seal which comes into contact with the cylinder wall. The central part of the piston forms a valve, in this case a rubber disc sitting over an aluminium plate in which several holes have been drilled. As the pump rod is raised and the piston valve moves up through the cylinder, the valve is closed and water is drawn up above the seal. This action causes the lower footvalve to open allowing water to enter from the well into the lower end of the cylinder. The water raised above the piston passes up the rising main (drop pipe) and spills out at the surface through the outlet pipe. When the piston is lowered through the cylinder, the footvalve closes and water passes through the piston valve itself. On this stroke, the piston moves through the water which remains relatively static. On the next upstroke a further charge of water is lifted above the piston valve to the surface — hence the name 'lift pump'.

Pumps fitted with buoyant pushrods like the Nsimbi, appear to deliver water on the downstroke. This is because the large volume pushrod displaces water in the drop pipe as it moves downwards. This is much the same as stepping into a bath completely full of water — the water spilled over the side has the same volume as that part of the body immersed in the bath. The same volume of water is lost on the up stroke, as the rod moves out of the water.

Installation of the Nsimbi Pump

The well cover should be cast with the studs of the pump head embedded in it. The well cover is cemented to the well and the apron and water run-off channel are made. The water delivery pipe is taken off the pump head just above slab level, so the well head must be raised, and the apron extended to allow buckets to be fitted under the outlet pipe.

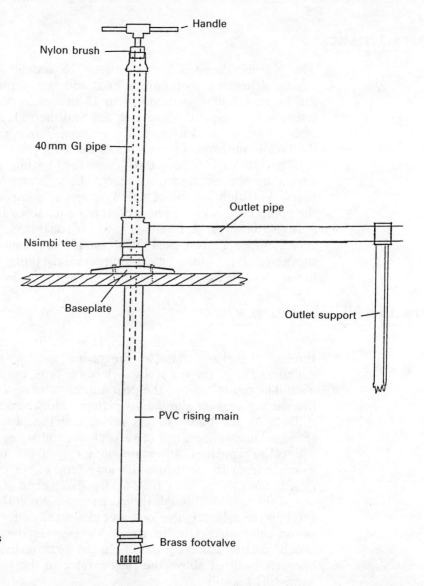

The Nsimbi Pump was Designed by Klass Jellema of the LWF.

Nsimbi Pump — Mark II.

Cutting and joining pipes

The well is measured for depth, and the pump rods and rising mains cut so that the footvalve lies 500 mm above the well bottom. The brass footvalve is fitted to the adaptor on the lower end of the PVC cylinder, which is a metre long. The PVC cylinder is cemented to the rest of the rising main through a 50 mm PVC socket. The head of the rising main is fitted with a PVC collar which is held up by the base plate. The cylinder and rising main are laid on the ground and the rods cut to match for length. The plunger is cemented to the lower end of the rod through a special adaptor. The upper steel section of the pushrod, to which the handle is attached, is cemented to the upper end of the rod also through a similar adaptor.

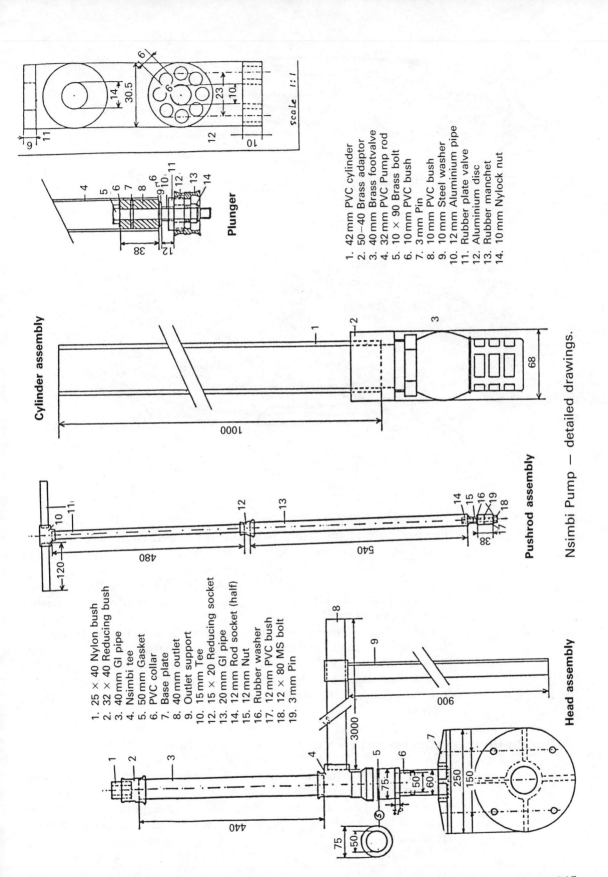

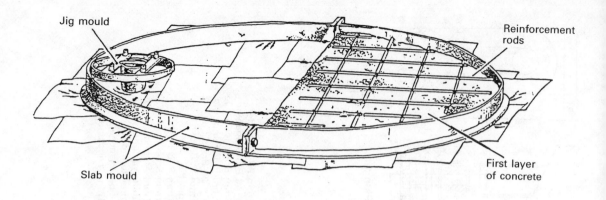

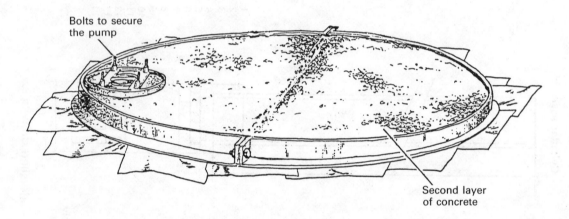

Fitting the pipes to the well

The baseplate of the pump is bolted to the well cover, and the rising main lowered through the plate so that the PVC collar rests inside the plate.

The pushrod assembly is then cleaned, and the tee handle unscrewed from the pushrod. The entire pushrod with piston attached is pushed down the rising main.

The rest of the pump head is now screwed on to the baseplate with the baseplate gasket in place. The pushrod is passed through the nylon bush and the tee handle screwed back on to the rod. The 40 mm outlet pipe is then added and the final position adjusted to suit the well head. It is always wise to support the water outlet pipe with an outlet support.

Maintenance of the Nsimbi Pump

Like all handpumps the Nsimbi will require maintenance and repair, and these in turn must be supported by a maintenance system supplied with suitable tools and spare parts. The Nsimbi pump, like the Blair Pump cannot easily be repaired without specialised knowledge and spare parts.

It is wise to inspect the following parts for looseness and wear.

1. Baseplate and pump head

Check the baseplate retaining bolts for tightness and ensure that the pump head is threaded up tight against the baseplate. Any looseness at this point will result in leakiness and potential contamination of the well. The threaded end of the pump head where it screws into the baseplate may wear badly in some conditions, and replacement parts should be available for this unit.

2. The handle guide

This is made of nylon and should have a life of several years. It is important to ensure that it is fitted tightly. It should be replaced when worn.

3. The piston valve

This is also known as the plunger, and is a reliable unit, but like all other parts is subject to wear. The diameter is originally about 41.5 mm and can wear back to 40 mm in two or three years when used by a few families. In hardworking settings replacement may be required every year. When this part wears, efficiency is lost and the seal must be replaced.

4. The footvalve

This is a standard unit usually made of brass and is subject to the same stresses as in other lift pumps. It is wise to use well designed and engineered units, as these will continue to function well for years. Poorly designed and manufactured units can give many problems. The rising main must be extracted to examine the footvalve. Footvalves can jam open with grit which is sucked into the lower end of the cylinder. A good stainless steel screen helps to overcome problems of this type.

5. Steel to PVC pushrod adaptors

These connect the steel section of the pushrod (and handle) and the plunger to the PVC section of the pushrod. They may unscrew or break depending on the amount of use they have. The buoyancy of the pushrod is also dependent on a good water-tight seal being made at this point. Spare adaptors should always be available in a spares kit. If an adaptor breaks, this should be unscrewed from the handle or valve and cut off from the PVC pushrod. A new unit should then be cemented to the pushrod and screwed into the handle or valve.

6. Cylinder, rising main and pushrod

These may fracture during use, but can be repaired easily if PVC cement and PVC pipe sockets are available. A stock of 50 mm and 32 mm sockets should be placed in the maintenance kit. The broken ends should be cut off straight, cleaned and reconnected with the new socket.

It is also wise to keep a few 50 mm PVC collars, which are cemented on to the top of the rising main. The rising main is supported by the collar within the baseplate, and should this break, the rising main section of the pump will fall into the well.

The cylinder is a one metre length of 50 mm class 16 PVC pipe. After some years of use, the internal diameter will increase and the efficiency of the pump will be reduced. Spare lengths of pipe should therefore be stocked to replace worn cylinders.

Making a direct action pump

The Nsimbi is one of a number of direct action pumps which are used in various parts of the world. These pumps use sound principles and are popular when they are in use. Very often however, the pump is let down by the use of weak materials which do not suit community conditions. A community pump should be very robust, so that it is able to withstand a great deal of heavy use. For lighter duty settings PVC piping which is easy to cut and joined can be used successfully.

1. Lighter duty models

These can be fabricated in the same way as the Nsimbi Pump with 50 mm class 16 pipe as the cylinder/rising main and 32 mm class 16 pipe as the pushrod. If a hand made pump is fabricated, special attention should be paid to three parts: the plunger, the pushrod adaptors and the pump head which supports the rising main from the well head. All these parts may require specialised fittings.

The pump head

One way of making this part is to thread a 50 mm PVC thread to pipe adaptor into a 50 mm GI socket and embed these in a concrete headblock 150 mm thick. This will leave approximately 20 mm of GI socket protruding above the concrete. The internal diameter of the adaptor is 64 mm and a PVC sleeve 50 mm long must be cemented within the adaptor to reduce the diameter to 50 mm. The 50 mm rising main can be cemented into this at a later stage.

The plunger

This can be made in rubber so that it can act as both a piston seal and a non-return valve. A thick rubber disc should be cut to fit neatly within a 50 mm class 16 pipe (diameter 41 mm) to form the seal. The same disc should be drilled with holes to form the valve seat. The valve itself should also be made of a rubber disc. The valve shaft can be made from a 10 mm nut and bolt approximately 90 mm long. A suggested design is illustrated in the figure.

Nsimbi Pump — suggested design for light duty model.

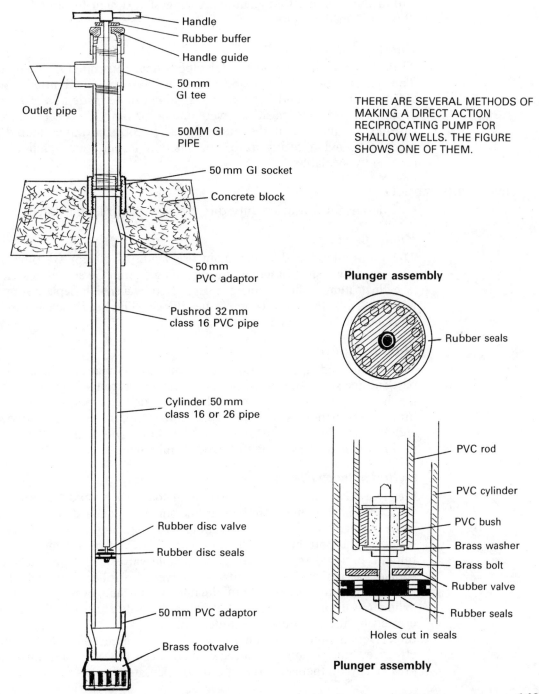

THERE ARE SEVERAL METHODS OF MAKING A DIRECT ACTION RECIPROCATING PUMP FOR SHALLOW WELLS. THE FIGURE SHOWS ONE OF THEM.

Steel to PVC pushrod adaptor

The handle and upper section of the pushrod is made of steel and the main pushrod of PVC. This adaptor is used at the steel/PVC joint. The adaptor is based on a 12 mm nut and bolt 80 mm long. The nut and bolt are tightened around a 40 mm long PVC bush with an inner diameter of 12 mm and an outer diameter of 24 mm. The threaded end of the bolt of the adaptor is screwed into a half 12 mm rod socket which is welded to the upper steel pushrod, the PVC end being cemented within the 32 mm PVC pushrod. A suggested design for an adaptor is illustrated in the figure.

Pump head

This can be made from 50 mm GI pipe fittings illustrated in the figure. The outlet pipe can be made about 300 mm long and fitted about 500 mm off the ground. The pushrod guide can be made of a 50 mm to 25 mm GI reducing bush screwed into a 50 mm GI socket which is threaded to the top of the pump head. Ideally the guide should be made of a nylon fitting as in the Nsimbi Pump, but nylon is not commonly available.

2. Heavy duty model

One design for a heavier duty direct action pump is illustrated.

Pump head

The pump stand and rising main are made of 50 mm GI pipe and supported by a steel baseplate into which a 50 mm GI socket is welded (see illustration). The baseplate is supported by a steel baseplate support which is embedded in a concrete headblock. The headblock is cement mortared on to the well slab.

Handle/pushrod

The handle and upper section of the pushrod is made of 20 mm steel bar, and tee shaped at the upper end. It passes through a guide/bush at the head of the pump stand, which in existing models is made of bronze. The handle is equipped with rubber buffers at upper and lower ends, mounted on the pushrod itself. The lower and main section of the pushrod consists of a string of 12 mm mild steel rods.

Cylinder-footvalve

Ideally a 40 mm cylinder with matching footvalve should be used, but these are uncommon in Zimbabwe, the smallest standard being 50 mm. Thus normally a 50 mm brass cylinder with matching brass footvalve is fitted on the bottom of the rising main. This can be a standard type, but the prefered type has fully extractable valves. The standard seals should be replaced with suitably machined rubber discs which fit neatly inside the cylinder, but do not expand and press hard on the cylinder wall on the upstroke. If standard seals are used, the friction created by the seal on the cylinder wall makes pumping harder in the direct action pumps. When rubber discs are used, the seal is not so effective, but friction is reduced and rods can move up and down at a greater speed. Increased rod speed compensates for poorer seals.

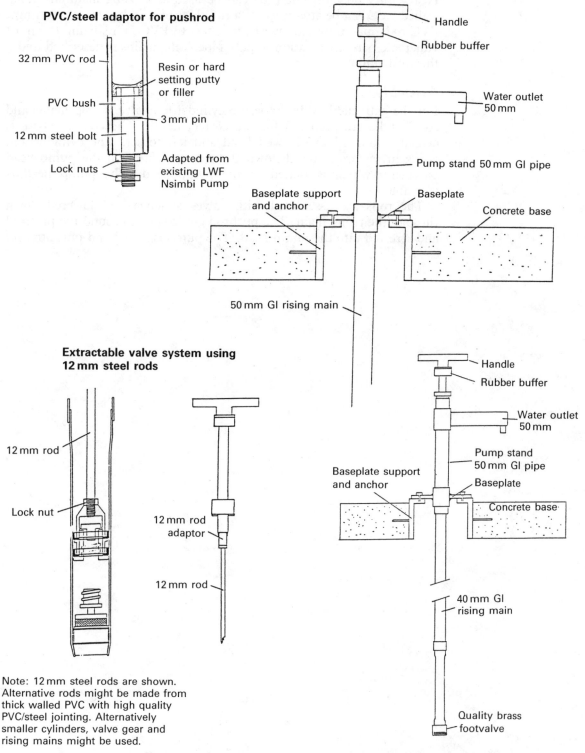

Nsimbi Pump — suggested design for heavy duty model.

There are several methods of making a direct action reciprocating pump for shallow wells. The figure shows one of them.

Alternative rising mains and pushrods

Heavy duty polyethylene pipe can be used as a rising main and thick walled PVC can be attempted as a rod. However in heavy duty settings, PVC rods may prove too weak, the steel to PVC joints being the most vunerable part. Installation of poly pipe rising mains is described under the Bush Pump.

Assembly

For the extractable valve type, the cylinder/footvalve, rising mains and rods/handle are prepared for the correct depth. The concrete base is cement mortared to the well head and left to set. The cylinder and rising mains are lowered down the well and fitted to the pump head baseplate, which is bolted to the pump head. The pump head is assembled.

The rods are assembled with valves attached and lowered down through the rising main. The pushrod guide is fitted round the pushrod and screwed into the pump head. The pump is tested and put into use.

The Bush Pump

The 'Bush Pump' is the most successful and most important handpump used in Zimbabwe, and has been in service since 1933 when it was first designed by Tommy Murgatroyd, a Water Supply Officer operating in the Plumtree District of Matabeleland. Its success can be judged by the fact that no other hand operated pump, save the bucket and windlass, is more common in the communal lands of the country. It remains the pump of choice for all deep well and borehole settings and in all heavy duty settings, even when they draw water from shallow aquifers. Currently about 15,000 Bush Pumps are in service throughout Zimbabwe.

The Bush Pump has been restyled several times since 1933 when Murgatroyd performed his pioneering work. The first major change took place in the mid 1970's when Cecil Anderson redesigned it and called it the 'Bush Pump', which became the new Government Standard Handpump. Following Independence, in 1980, a number of foreign organisations attempted to introduced their own handpumps into Zimbabwe, but this was strongly resisted by the Government. The

The Bush Pump is the most commonly used handpump in Zimbabwe. This is a model 'A' type installed over a well.

The Zimbabwe Bush Pump ('B' Type).

period after 1980 was also characterised by a proliferation of Bush Pump models, with several being inferior in design or manufacture to the standard, but some having improved features. In 1987, the Government designed and introduced a new standard Bush Pump and also recommended that further research should be carried out on the Bush Pump design in order to lower the costs and complexity of maintenance. As a result a new Bush Pump was introduced as a standard in 1989.

The fact that the Bush pump has survived for so long strongly supports the view that the basic concept is very sound. The pump owes much of its success to three features specified by Murgatroyd, namely simplicity, durability and ease of maintenance. It is a very forgiving pump, and is able to endure much punishment yet will still perform when many parts are badly worn out. Nothing is hidden from view in the pump head, which remains accessible for ease of maintenance and repair. The use of a hardwood block, which operates as both a bearing and lever system, is perhaps the key feature of the Bush Pump. The teak block is robust, durable and self lubricating and still operates when very old and badly worn. These features have made the Bush Pump one of the most successful and longest used handpumps in the Developing World.

Description

The Bush Pump operates on a lift pump principle, the reciprocating action being transferred from the pump head to the cylinder through a series of galvanised steel pump rods running inside a steel pipe (rising main). Most rising mains are made from 50 mm galvanised iron pipe, although 40 mm pipe is becoming more common. Most rods are made of 16 mm mild steel although 12 mm is also used. Pump cylinders are made of brass and are either 50 mm or 75 mm in diameter. The piston and footvalves are also made of brass. Most piston valve seals are made of leather, but neoprene is becoming more common. Several methods of easing the burden of maintenance are being developed, especially for 'down the hole' components. These include a fully extractable piston and footvalve system made of brass for use with 50 mm steel rising mains. Heavy duty polyethylene pipe is also being used on a small scale together with drawn 12 mm steel (EN 8) rods, which with care can be lowered and raised in one piece within the pipe.

The pump head

Most Bush Pumps installed in recent times consist of a steel pump stand bolted to the upper end of a steel borehole casing (see illustrations on following pages). The original Murgatroyd Pump was set into a concrete base, and some later models also used this feature. The new standard model is designed to bolt on to a steel borehole casing and, where the pump is fitted to a well, a length of borehole casing is cast into the concrete well cover. The use of large 16 mm U bolts which clamp the pump to the casing is a very successful technique.

All Bush Pumps use a durable hardwood block as a lever and bearing surface — a system that has been supremely successful for over half a century. This is the model 'B' Bush Pump.

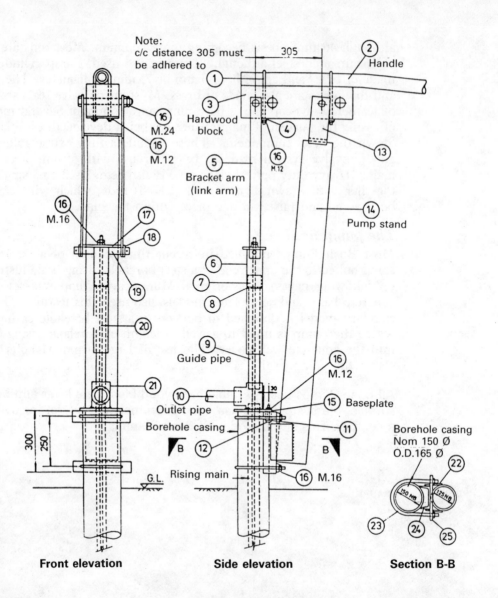

Front elevation   **Side elevation**   **Section B-B**

List of Parts

Item	No.	Material
1	2	12 mm Ø U-bolt 252 × 72 c/c
2	1	50 mm GI pipe 3000 length
3	1	Hardwood block 150 × 150 × 500
4	2	6 mm M.S. Flat 150 × 40
5	2	10 mm M.S. Flat 700 × 40
6	1	65 mm Black pipe 450 length
7	1	60 mm Ø rubber buffer 50 length
8	1	60 mm Ø Steel washer 3 thick
9	1	50 mm GI pipe 450 length
10	1	50 mm GI pipe 430 length
11	2	6 mm M.S. Flat 30 × 30
12	2	12 mm Ø Stud and steel washer 85 length
13	2	16 mm M.S. Flat 210 × 80
14	1	125 mm Black pipe 1800 length
15	1	10 mm M.S. Flat 200 × 200
16	6	M.12 Nuts
16	6	M.16 Nuts
16	4	M.24 Nuts
17	1	10 mm M.S. Flat 165 × 100
18	2	10 mm M.S. Flat 50 × 50
19	2 Each	24 mm Ø M.S. Rod 90 length, 3 mm Ø Split pin
20	1	16 mm Ø Rising rod as req'd
21	1	50 mm Std. Black tee
22	2	10 mm M.S. Flat 100 × 100
23	2	16 Ø U-bolt 269 × 182 c/c
24	1	127 × 64 Channel 300 length
25	2	16 mm M.S. Flat 250 × 50

The old standard Bush Pump.
The most common handpump in Zimbabwe.

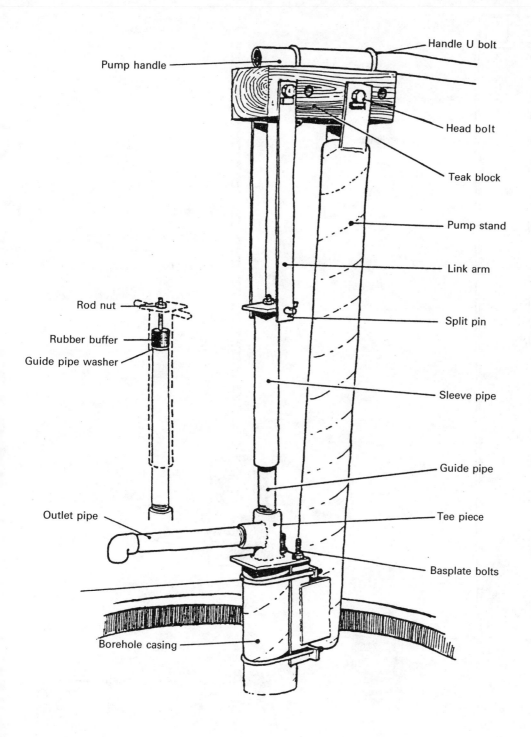

Old Standard Bush Pump — Parts.

This model is the most common and familiar handpump in Zimbabwe.

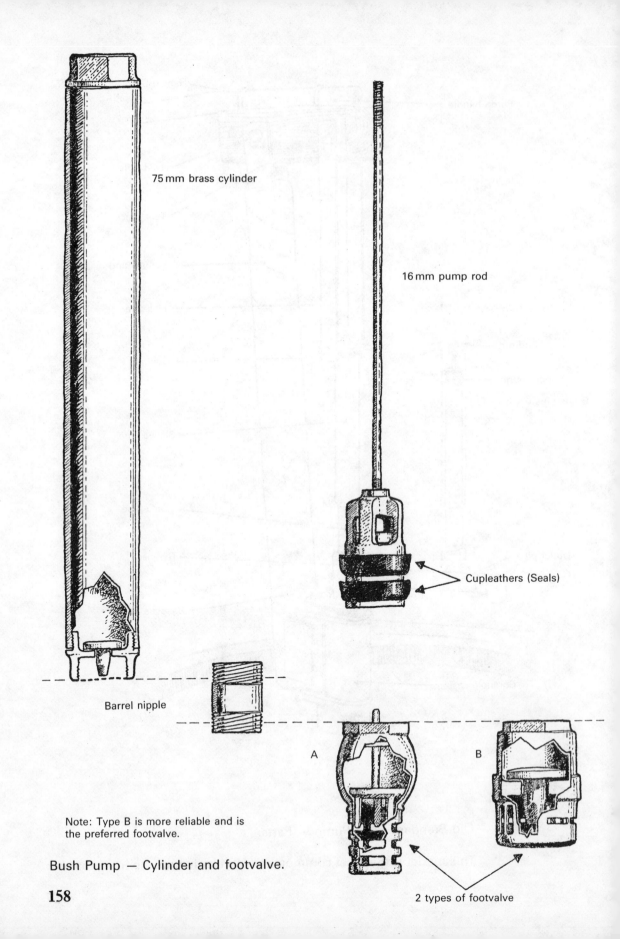

Bush Pump — Cylinder and footvalve.

Note: Type B is more reliable and is the preferred footvalve.

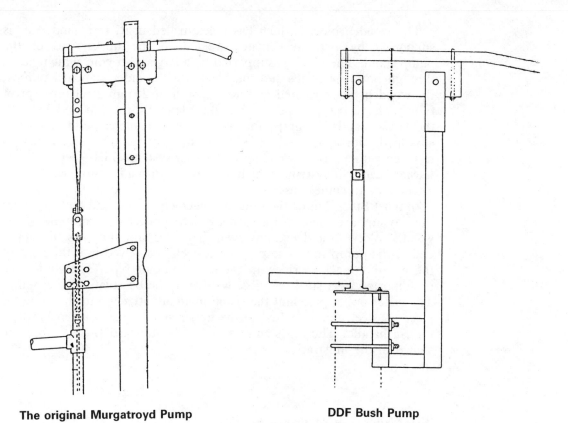

The original Murgatroyd Pump

DDF Bush Pump

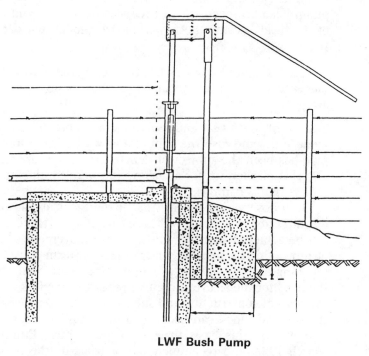

LWF Bush Pump

Types of Bush Pump.

159

The wooden block, which has holes drilled in it, front and rear, is supported by two steel plates welded to the upper end of the pump stand. The block is supported by a large head bolt, which passes through both the plates and the block. In the older standard pumps, the wooden block rotated around a length of 25 mm steel pipe (pivot tube), which was clamped within the plates by the nut and bolt. In the latest standard pump, this is a 35 mm diameter solid steel bolt equipped with a squared head, to avoid rotation. The bolt is manufactured with a shoulder and spring washer system which keeps it tight. Earlier head bolt systems, which were fitted with a lock nut system had a tendency to come loose.

In most Bush Pumps the same arrangement of nut and bolt system is used to support a pair of link arms which connect the front end of the wooden block to a sleeve pipe, which guides and supports the upper end of the pump rod. The pump rod itself is screwed into the head of the sleeve pipe which rides up and down a guide pipe. The guide pipe is supported by a steel tee piece welded to the baseplate of the pump. The water outlet pipe and the rising main are attached to the same tee piece. A steel pipe is bolted to the upper side of the wooden block to form a handle. The old standard and 1987 standard Bush Pumps are shown in the illustrations.

Simplification of pump head

A smaller more compact unit, called the Model B, which retains the strength and durability of the old standard model, but simplifies the pump head arrangement and reduces the capital and maintenance costs of the head became the new standard model in 1989.

It was once said that:

"The designer knows he has reached perfection, not when there is no longer anything to add, but when there is no longer anything to take away."

The problem of wear and replacement of the link arms, sleeve pipe and guide pipe, and also the rubber and washer assembly inside the sleeve pipe has been the subject of debate for some years. It has been argued that this mechanism for attaching the wooden block to the pump rod is unnecessarily complex.

Close observation reveals that if the block is allowed to rotate equidistantly above and below the horizontal, a stroke of at least 230 mm can be achieved with very little lateral movement of the rod, less than 50 mm. This means that the rod can perform a useful stroke without touching the side walls of a 50 mm rising main. Such a system eliminates the need for the sleeve and guide pipes and the link arms, thus reducing the costs of construction and subsequent maintenance. The link between the rod and block can become direct, simpler and thus more effective.

It was on this basis that the Type B Bush Pump was designed in March 1987, and recommended for trials by the NAC. In 1989 it was recommended for national use. The Type B is simpler and more compact

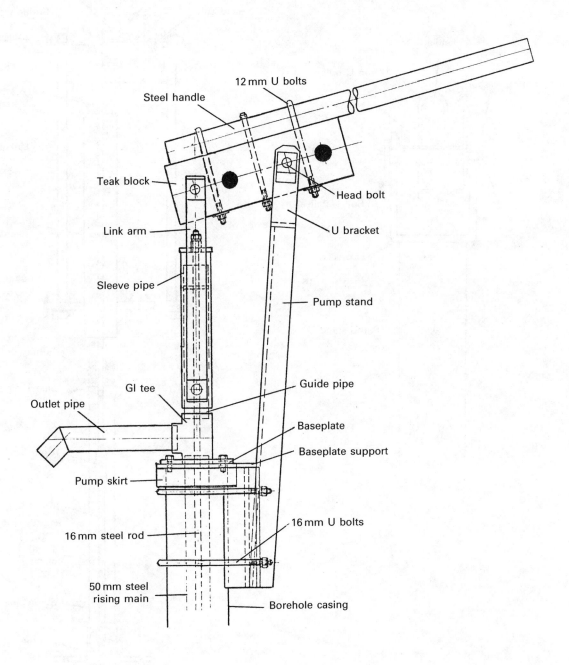

'A' Type Bush Pump.

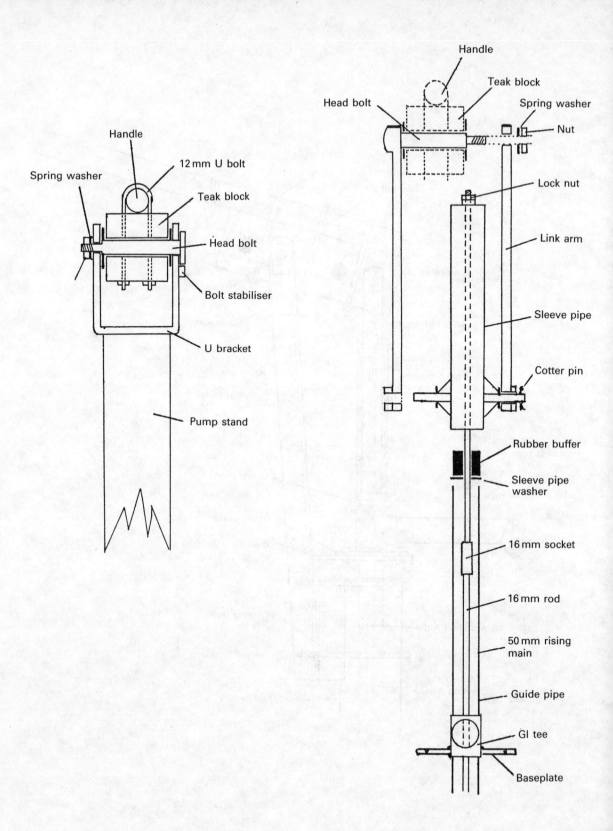

Detail of 'A' Type Bush Pump head assembly.

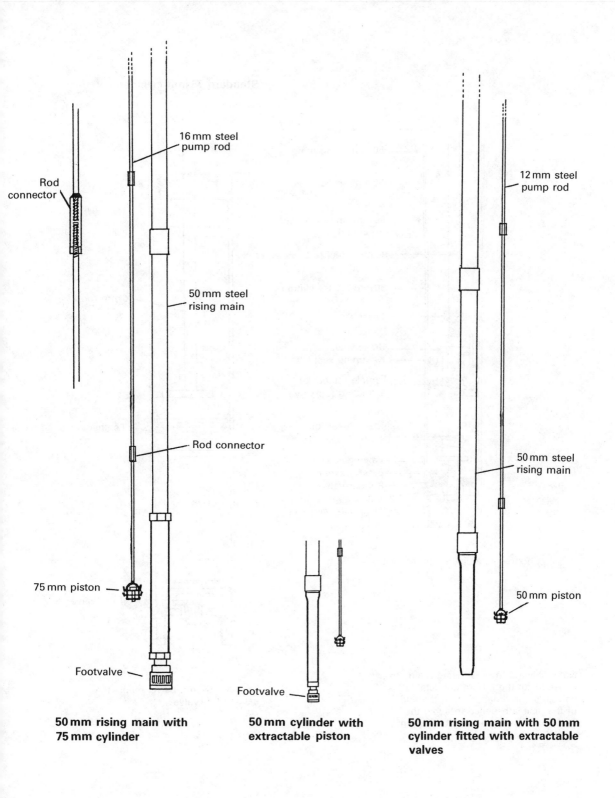

Bush Pump rods and rising mains.

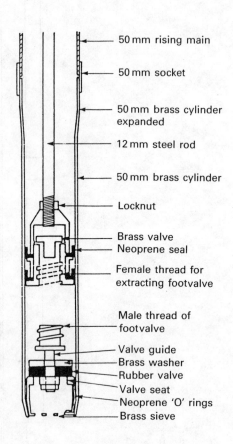

Extractable piston and footvalve

Note: When 50 mm GI rising main is used with the 50 mm extractable valves, the internal edge, at the end of the pipe is bevelled to avoid the seal catching on the pipe at this point.

Standard 75 mm cylinder

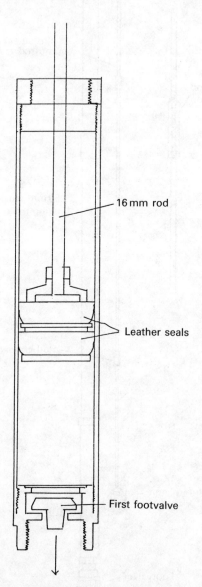

Through 50 mm barrel nipple to second footvalve

Bush Pump cylinders.

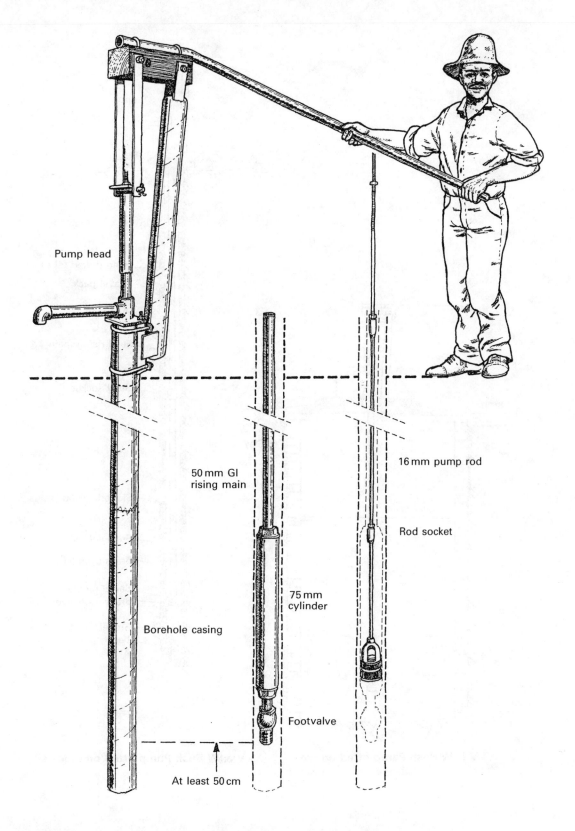

Bush Pump. Arrangement of pump and rising main in a borehole.

165

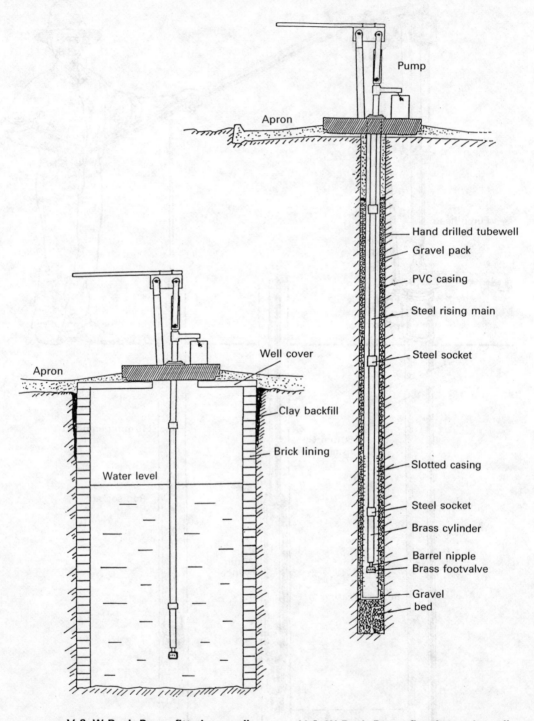

V & W Bush Pump fitted on well   **V & W Bush Pump fitted on tubewell**

Bush Pumps on wells and tubewells.

than the Type A and consequently cheaper and easier to fabricate and maintain. It retains all the sound features of the Bush Pump, namely the hardwood block, steel pump stand, U bolt clamping system, solid pivot bolt system, etc; but has less moving parts to wear, and the stability of the rod is maintained throughout the full stroke. Whilst retaining the strength of the original Bush Pump, the Type B is far more compact which makes handling easier and fabrication simpler.

The pump rod passes through a 'floating washer' system which is able to accommodate all horizontal movements of the rod; forward, backward and lateral. Two floating washers are used in the floating washer housing, the upper unit as a support for the rubber buffer, the lower washer as a hygienic seal. Bacteriological tests carried out over a two year period show that water quality is not impaired by this system.

The main wearing parts in the head are reduced to the wooden block and the washers. The upper washer wears faster than the lower, since it has a heavier load to bear, but present results show that washer replacement need not be frequent. The washers are 100 mm diameter discs of 6 mm thick steel with a central 17 mm hole drilled in them. They are cheap to make and easy to replace. The wooden block lasts for many years and is drilled with two sets of holes which doubles its life, which can extend to twenty or thirty years. The pump is supplied with its own spanners and a set of spare washers. The pump is fully illustrated in the figures.

Characteristics

Bush Pumps are installed in heavy duty communal settings throughout Zimbabwe, where they may serve several hundred people. In many cases they are kept working for most of the day.

Bush Pumps can lift water as high as 100 metres, but most are installed to depths between 30 and 40 metres. The pump is very flexible however and will also operate effectively on shallow wells. Most Bush Pumps are fitted with 75 mm diameter cylinders, and under ideal conditions they can deliver up to 30–40 litres of water per minute. This discharge rate is reduced at greater depths, and also when the leather seals are worn or the footvalve leaks. 50 mm cylinders deliver between 15 and 20 litres per minute. The length of each stroke is very variable, but normally lies between 75 mm and 150 mm. The maximum stroke is about 220 mm. The down-stroke is buffered by a rubber block mounted around the rod, the up-stroke is controlled by the handle which meets the ground.

Several parts of the Bush Pump are subject to wear, more so in the older units than in the new standard model. For the older units these parts include the pivot tubes and bolts in the headblock assembly, the link arms, sleeve pipe, guide pipe and to a much lesser extent the wooden block. In all cases the piston seals wear at a rate which is dependant on the quality of the seal itself, the usage and the physical quality of the water. Water containing sand or sediments leads to greater rates of wear of the seals. Seals may require replacement every six months or once every two or three years. In some cases they appear to last for many years without replacement. 75 cm seals wear out less rapidly than 50 mm seals, since they are larger and travel a smaller distance for every litre of water delivered.

Wooden blocks should last for at least ten years and in many cases over twenty years before they wear to the point of being inoperative. Wooden blocks are drilled with two pairs of holes which double their working life. When the first pair is worn out, the block is moved and the second pair is put into use. The hardwood blocks are made of teak or mopane and after drilling they are boiled in oil for two hours and left to cool overnight. This process forces air out of the wood, and replaces it with oil. Thus the block contains its own lubricant.

Steel rising mains and rods also break and wear or simply corrode away. Rod separation may result from unscrewing joints or the rods snapping at the threaded joint. 16 mm rods are about three times as strong as 12 mm rods. Since both rods and rising mains are taken out, unscrewed and rejoined each time the piston seals are replaced, the threading of the joints is also subject to wear or damage. An increasing amount of attention is being paid to fully extractable valve systems, or rods and rising mains that do not require unthreading for extraction.

How it works

This is best explained by referring to the diagram on the next page. The pump operates on a 'lift pump' principle with the reciprocating action of the head being transferred to the piston and cylinder through a series of rods. The cylinder is attached to the rising main and the piston to the rod. The base of the cylinder is equipped with a non-return valve (footvalve) and may have two of these units. The piston is also equipped with a valve (piston valve) and a set of seals which make a water tight seal against the brass cylinder wall. The rising main is attached to the pump base and the rods to the wooden block, through the sleeve pipe and link arms or U bracket. As the wooden block is raised and lowered the piston moves up and down the cylinder. This action causes water to be lifted from the well, through the cylinder and rising main to the surface.

The lift pump mechanism

On the upstroke, the piston valve closes and the footvalve opens and the piston moves upwards through the cylinder lifting water above it through the rising main to the surface. At the same time fresh water is drawn into the base of the cylinder from the well or borehole through the open footvalve. Water is discharged at the surface through the water outlet pipe attached to the pump head. As soon as the upstroke is complete, the footvalve closes, and holds up the entire column of water held in the rising main.

On the downstroke, the piston moves downwards through the column of water held in the cylinder, its valve being open to allow water to pass through it. The footvalve is closed at this time, and in fact there is no movement of water through the rising main during this stroke. Water is lifted again when the piston passes upwards through the cylinder.

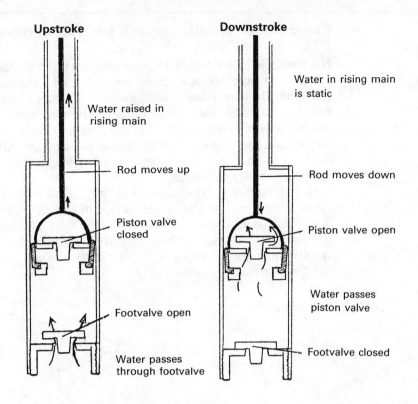

The operation is most effective with watertight seals and valves. If the seals are worn some water will slip by the seal as the piston rises and output will be reduced. Similarly, if the valves are leaky, water output will be reduced. A leaky footvalve allows water to be lost from the rising main and can reduce output considerably.

This is a very successful mechanism. Reciprocating pumps using similar principles have been used since Roman times. Almost all modern hand operated pumps use this principle. Where a principle is long lived and is widely used, it must have a great deal of merit.

Methods of installing the Bush Pump

The standard Bush Pump is designed for attachment to a length of 150 mm diameter borehole casing. Obviously if the pump is fitted to a borehole lined with this steel casing, the upper end will be chopped off, about 500 mm above ground level and the pump attached directly. In the case of a borehole or tubewell lined with PVC casing a length of steel casing will be embedded in a concrete anchor around the head of the borehole and the pump fitted to this. In the case of a wide diameter well, a 500 mm length of casing is embedded into the concrete well cover, with 175 mm embedded in the concrete itself and 350 mm left free for attachment of the pump head. These methods of preparing the water point for pump attachment are now described.

Preparing for installation on a borehole or tubewell lined with PVC

If a borehole is lined with 150 mm steel casing it is standard practice to cut off the pipe 500 mm above ground level. This will leave enough casing for the apron and in addition over 300 mm which is the minimum length suitable for pump attachment. The steel pipe should be cut off square, and it is important that the borehole is surrounded by a reinforced concrete apron at least 2 metres in diameter with a water run-off channel at least 6 metres long, which directs water to an adequate soakaway or drainage area.

In the case of a borehole or tubewell lined with PVC casing, it is essential to surround the upper section of the PVC casing with 700 mm of 150 mm steel casing for pump attachment. The lower 250 mm surrounds the PVC casing below ground, the upper 500 mm allows for an apron thickness of up to 125 mm above ground level and a further 325 mm for pump attachment.

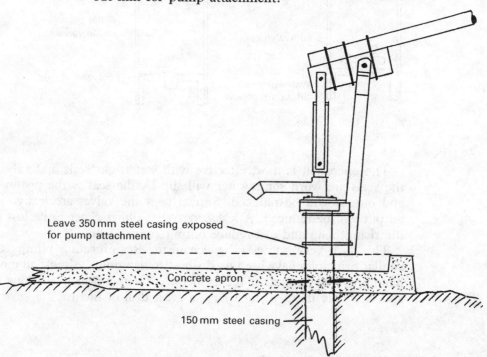

Fitting a Bush Pump on a borehole lined with a steel casing.

Where boreholes are lined with PVC casing, several metres of the annular space at the head of the borehole are filled with a concrete grout. In the case of tubewells this is normally one metre. Before the steel pipe is fitted it is essential to open out the diameter of the drilled borehole or tubewell from its existing 150 mm or 175 mm diameter to 500 mm diameter to a depth of 250 mm. The PVC casing is cut off 450 mm above ground level.

The 700 mm length of 150 mm steel casing is then placed around the PVC casing, so that 450 mm is above ground level and 250 mm lies

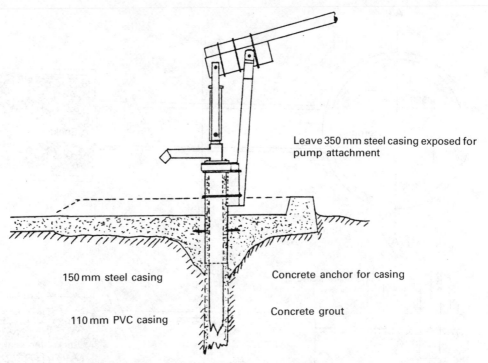

Leave 350 mm steel casing exposed for pump attachment

150 mm steel casing

110 mm PVC casing

Concrete anchor for casing

Concrete grout

Fitting a Bush Pump on a borehole or tubewell lined with a PVC casing.
In the case of a tubewell or borehole lined with PVC casing the pump is supported by the steel casing anchored in the concrete cast at the head of the water point. It is important that this anchor is made in a strong mixture of reinforced concrete, steel sprags welded to the casing add further support.

below it. The space between the two casings is then carefully filled with strong cement mortar, and the hole surrounding the steel casing is filled to ground level with a strong mixture of concrete (3 parts stone, 2 parts river sand, 1 part cement). This is left to set for two days.

The apron and water run-off channel (headworks)

This can be built the day after the casing is fitted as follows:

The apron and water run-off channel can be made with bricks and concrete or with a special mould designed for the purpose by V & W Engineering, Harare.

A hand built apron and water run-off built for a Bush Pump around a borehole or tubewell.

171

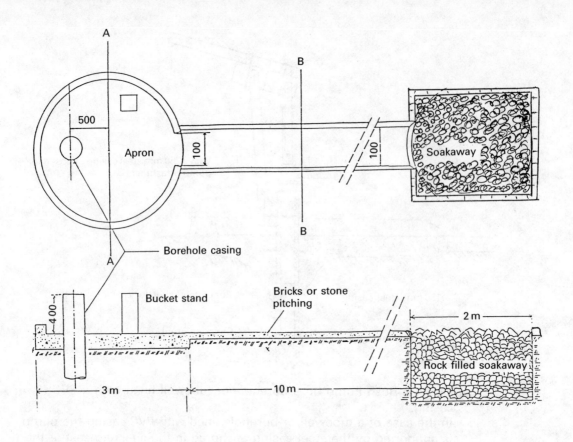

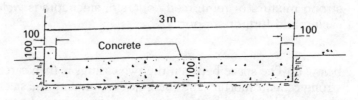

Section A — A

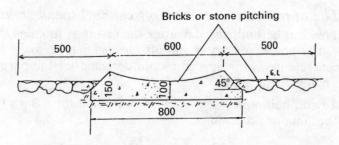

Section B — B

Plan and section of apron and water run-off made for a Bush Pump fitted to a borehole or tubewell.

If the mould is not available, a circle of bricks or rocks is laid around the borehole casing so that the borehole lies to the back of the apron as shown in the diagram. The rocks or bricks are formed in a circle 2 metres or preferably 3 metres in diameter. These form the elevated rim of the apron, and are mortared in position. The space between the rim of the apron and the casing is filled with concrete reinforced with 3 mm bar. It should be at least 100 mm thick and sloped to allow complete drainage into the water run-off. The apron should be well reinforced around the casing to add additional support. The apron should be finished off by steel floating.

The water run-off channel is made either with a mould or with bricks and concrete. This should be at least 6 metres long and 100 mm deep in reinforced concrete and shaped and sloped to ensure that all water drains away into a seepage area. The walls of the channel can be built up with bricks about 300 mm apart. Water run-off channels are well known for their ability to break up with subsequent ground movement, and they are best made with strong concrete reinforced with 3 mm steel bar.

The channel directs water into an area of seepage or a soakaway. Some care is required to design a suitable seepage area which will not become inoperative a year or two after construction. It is normal for large quantities of sand to travel from the apron along the channel into the soakaway area, and provision must be made for this.

A soakaway can be built 2 metres in diameter, a metre deep and filled with rocks, but this will eventually fill up with sand, and will require cleaning out. The rate of filling with sand will depend on the usage of the pump and the type of terrain surrounding the facility. Alternatively an area of seepage can be built to accept waste water and sand. This is a hollow in the ground, at least 3 metres in diameter with a central depth of 500 mm. The area can be planted with grass, bananas or other vegetation that will help to take water away from the soil. In areas where cattle or goats may destroy the vegetation, the area should be surrounded by a strong wire fence. If the facility is near a vegetable garden waste water can be directed into a place within the garden.

Headwork design with washing stand and cattle trough

Many Bush Pumps are installed in areas where cattle watering may be essential, and very often some distance from where people live. In these cases cattle troughs are often constructed near to the Bush Pump and connected to it with piping. Also washing slabs for washing cloths can be built near to the Bush Pump for the convenience of users. These steps are often essential and are described briefly below. More detailed descriptions of Bush Pump installation and headworks design and construction are available from the DDF and have been written by Mr David Williams and Mr Roland Haebler.

A typical layout for a complete headworks design for a Bush Pump is shown overleaf. The apron and water run-off itself is best made with specialised shuttering designed for the job and the cattle trough can also be made with shuttering. Where shuttering is not available, the apron, run-off, cattle trough and washing stands can be made with bricks and mortar.

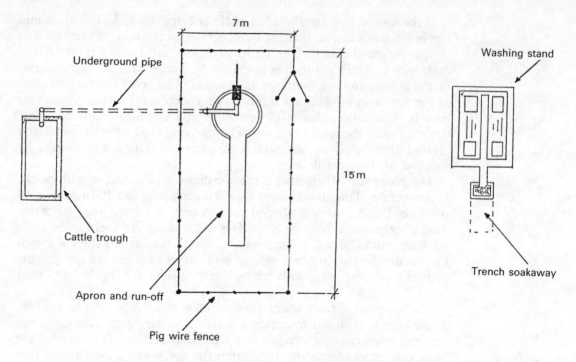

Arrangement of Bush Pump headworks.

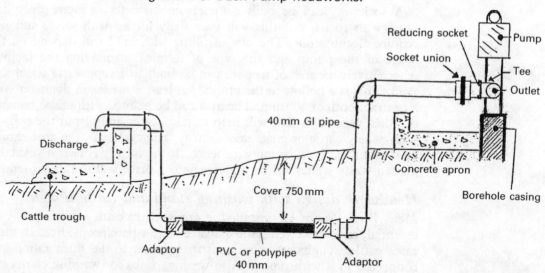

Arrangement of piping to cattle trough.

A fence can also be constructed around the apron and run-off channel. This is best made with stout gum poles and pig wire, reinforced with barbed wire.

The cattle trough is connected by a pipe to the handpump so that some water is delivered to the trough during routine use of the pump. When the pump is used to directly fill the trough, the end of the water outlet pipe can be closed off with a hand or with a plug made of wood or some other locally available material. Above ground components of this pipe must be made of steel, below ground components can be made with polypipe or PVC. Suitable adaptors will be necessary to convert from steel to the plastic pipe. 40 mm pipe is used and a 40 to 50 mm reducing socket is used on a 50 mm GI tee piece attached to the pump outlet pipe to tap off water for the trough. The top of the trough must be located downhill from the outlet of the pump and about 20 metres away. Three × 6 metre lengths of plastic pipe should be laid underground.

The water run off must drain away downhill and for the best orientation of the whole headworks, the apron shutter must be placed in position first. Once this is done, the fence, washing stand and cattle trough can be located. A suitable soakaway or garden should be placed at the end of the run-off channel to absorb waste water.

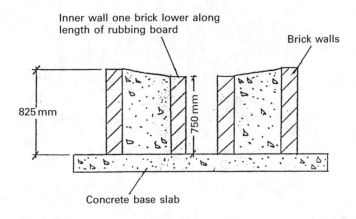

Cross-section of washing slab.

Preparing for an installation on a well

Bush Pumps have been fitted on to wells for many years in Zimbabwe, and this technique works well even on very shallow wells which deliver enough water throughout the year.

Where a new well is being constructed or an older well is being upgraded it is best to make a well cover slab with the 150 mm steel casing fitted into it. This is a well-established technique and is known to work satisfactorily. When a well already has a sound slab fitted to it, the slab can either be removed and a new one built for the purpose, or an extra slab can be built to fit above the existing one.

Making a well cover slab fitted with a casing

This must be made in reinforced concrete to fit the specific well, and will normally be about 1.5 metres in diameter. Formerly, when well cover slabs were designed to support a length of steel casing they were made 150 mm thick. These were quite heavy to handle, however, and subsequent work has shown that thinner, well-reinforced concrete slabs can adequately support a casing if the section surrounding the casing itself is thickened up. The thickened section of slab can either be cast above or below the main slab, the latter technique being the neatest

Plan of washing slab

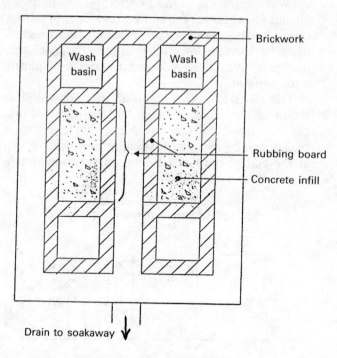

Plan of fencing

Cross-section of fencing

Details of headworks.

method. The main slab is 100 mm thick with the thickened section around the casing having a total thickness of 175 mm. One advantage of fitting a Bush Pump over a well is that the well cover can be made with a bucket access hole fitted with a separate concrete cover, mortared in position on the main cover slab. If the pump breaks down, the cover plate can be knocked off, and buckets can be used to raise water from the well. However, this is not always the preferred method and sometimes the well cover is made with no bucket access hole.

Technique

Mark off a piece of level ground with a circle of the same diameter as the final well cover. This will normally be 1.5 metres. A circle of bricks can be used as a mould, and these are placed around the circle, standing on edge.

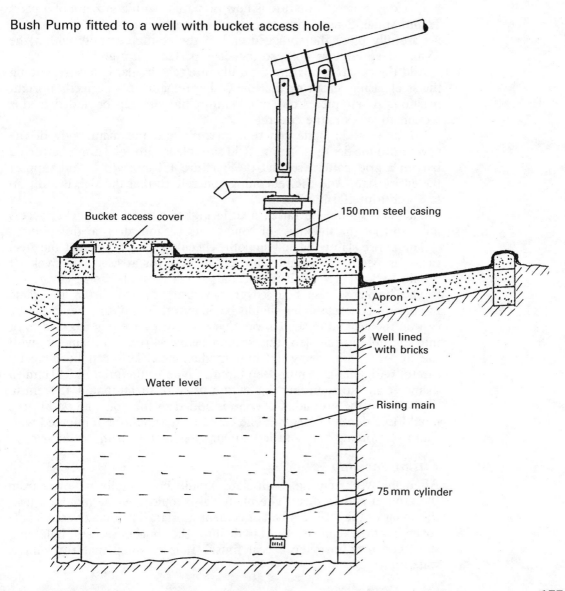

Bush Pump fitted to a well with bucket access hole.

A circular hole, 450 mm in diameter and 75 mm deep is dug with slightly tapered sides 150 mm from the edge of the slab as shown in the diagram. Before concrete is poured to make the slab, a layer of plastic or paper is laid over the ground both within the bricks and inside the circular hole. This will make removal of the final cover much easier. If plastic and paper is not available, a layer of river sand will help.

The length of the 150 mm steel casing should be 500 mm and a few lengths of steel bar (sprags) 300 mm long should be welded on to the lower 175 mm section as shown in the diagram. These will help to anchor the casing into the concrete.

Before the concrete is mixed and poured a circular mould 350 mm in diameter and 125 mm deep is placed within the main mould to form the bucket access hole. This is placed 150 mm from the edge of the slab as shown in the diagram. It is removed after one day, once the concrete has set.

A strong concrete mixture is now prepared, with a mixture of 3 parts 12 mm stone, 2 parts river sand and 1 part cement.

Carefully place the spragged end of the casing centrally within the 75 mm deep circular hole, so that it is perfectly upright.

Add the concrete mixture carefully into the circular hole surrounding the steel casing, ensuring that the casing remains in a perfectly upright position. A few pieces of 3 mm reinforcing wire can be added to this section to support the concrete.

Continue adding the concrete mixture into the main body of the cover until the depth is 50 mm. Add suitable lengths of 3 mm reinforcing bar, in a grid pattern so that the bars are 150 mm apart, and support the entire slab. Add the remaining concrete so that the slab is built up to a depth of 100 mm.

The slab should be raised a little higher around the bucket access hole, so that the thickness of concrete is 125 mm deep at this point.

The entire slab is floated smooth, checking once again that the steel casing is perfectly upright, and left to cure for at least one week. It should be kept wet at all times, and be placed under some cover.

The bucket access hole cover can also be made from the same mixture of concrete. This should be 50 mm thick. With care, it can be cast on the ground in such a way that the lower 25 mm is recessed to fit into the hole in the slab, this section being 340 mm in diameter, with the upper 25 mm being 450 mm in diameter. This can be done by carefully digging a 25 mm deep circular hole in the ground 340 mm in diameter and surrounding this with a brick or metal mould 450 mm in diameter on the ground. Concrete is added to the hole and within the mould to a depth of 50 mm. The cover plate should be reinforced with 3 mm reinforcing bar, and left to cure together with the main slab.

Fitting the slab

After one weeks curing, both slabs should be carefully removed from the ground and washed. The main slab should then be mounted over the collar of the well in a bed of cement mortar. The bucket access hole cover is then cement mortared into the bucket access hole and smoothed over to make a neat finish. It is essential that the joint is watertight.

The steel casing should now protrude 325 mm above the well cover, ready for pump fitting. The apron and water run-off channel are now made in a similar way to the borehole installation, although with a well, it is preferable that the diameter of the apron is 3 metres, allowing 0.75 m on both sides of the well head.

Note: Bucket access hole. When a Bush Pump is fitted to a well it is a wise precaution to fit a bucket access hole and cover which can be used if the pump breaks down. However great care is required to ensure that the cover is well mortared in position over the well cover. If it leaks, contaminated water will enter the well, and the advantage of fitting the pump will be lost.

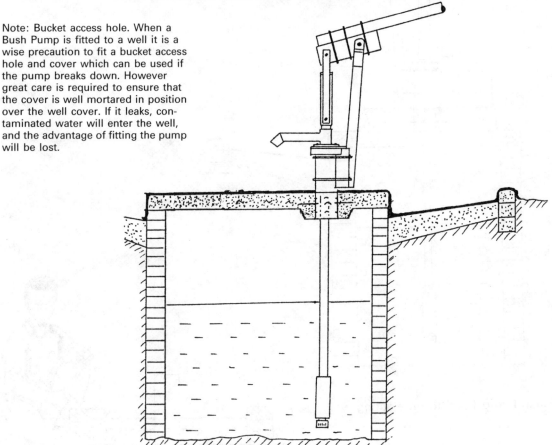

Bush Pump fitted to a well without a bucket access hole.

Making a casing slab to fit over an existing well cover

In some cases it may be more convenient to cast the steel casing into a separate slab which is mortared in position over the main slab of an existing well.

A piece of ground is levelled and a brick mould is built up 1 metre long, 0.5 metres wide and 175 mm deep. Two or three layers of bricks will be required for this mould. The steel casing should be 500 mm long and fitted with sprags as before. This casing is inserted into the middle of the mould so that it is perfectly vertical. The mould is now filled with a mixture of concrete made from 3 parts gravel, 2 parts river sand and 1 part cement, with pieces of 3 mm reinforcing bar added for strength. This is left to cure for one week, and is then mounted centrally over the well cover in cement mortar directly over the existing hole. An apron and water run-off channel are then built around the well.

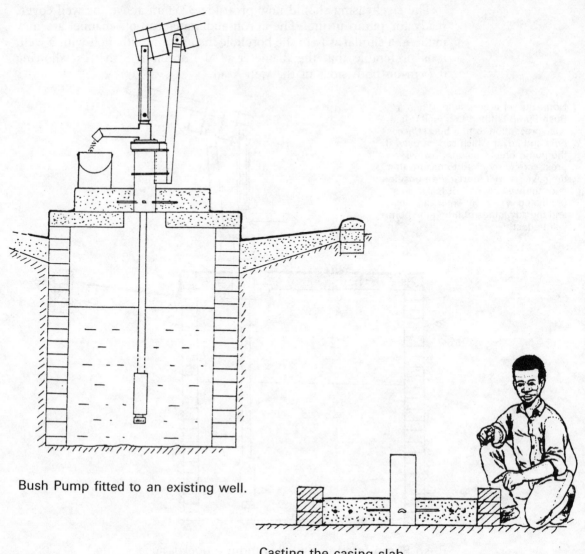

Bush Pump fitted to an existing well.

Casting the casing slab.

Fitting the Bush Pump

The method of fitting the Bush Pump is similar whether it is fitted to a borehole or to a well. Several models of Bush Pumps are in use throughout Zimbabwe, but they all have features in common. A few general points should be raised in connection with fitting all Bush Pumps. These can be listed as follows.

1. It is impossible to install a Bush Pump without the right tools. Spanners, pipe wrenches, a wire brush, a half round file, and a tin of plumbers' paste are some of the most important equipment essential for pump installation.
2. Before assembly all working parts should be checked and thoroughly cleaned.

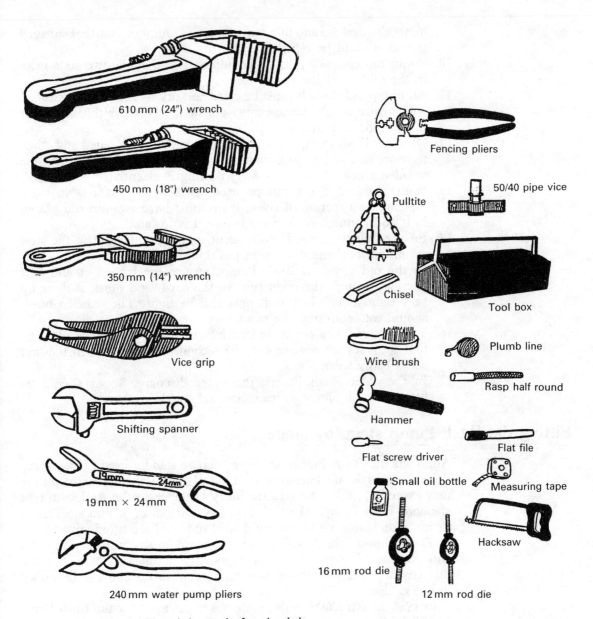

The right tools for the job.

   3. All nuts and bolts and joints should be tight. Far too many Bush Pumps fail simply because they fall apart.
   4. Clean and inspect all galvanised iron threads.
   5. Make sure the footvalve seats properly before installation
   6. Handle the cylinder with care — it is made of brass and can easily become distorted if turned with a pipe wrench.
   7. Make sure the piston components are clean and tightly assembled and the leathers are in good order.
   8. Clean the inside of the cylinder with a wet cloth before it is assembled. When reassembling the cylinder, use the hexagonal end caps.
   9. All rising mains should be inspected and cleaned and the threaded

joints also cleaned and fitted together with plumbers' paste. Damaged threads should be refurbished.

10. Pump rod connectors must be adjusted so that the two rods meet in the middle of the connector.
11. All pump rod threads must be clean and dry — they can be cleaned with a wire brush. Grease must not be used on any threads either in the rods or rising mains.
12. On older Bush Pumps, the pivot tubes should be cleaned and when they are assembled they should be able to rotate freely within the wooden block. A wood on steel bearing is essential.
13. With standard Bush Pumps used without the extractable valve system, each section of rising main must have a pump rod placed in it and connected properly before it is lowered.
14. Ensure that the large U bolts securing the pump head to the steel casing are horizontal when in position.
15. In the old standard Bush Pump, ensure the large nut and bolt passing through the pivot tube in the hardwood block is done up very tightly. The lock nut must also be tight. The wooden block should rotate around the pivot tube.
16. Two sets of holes are made in the Bush Pump wooden block. The first set is for immediate use, the second set for use when the first set is badly worn.
17. In the older Bush Pump, the water discharge spout should be fitted with an elbow so water is discharged downwards.

Fitting the Bush Pump stage by stage

Although the Bush Pump has now been re-standardised (1989) many different models are currently in use and the total number of pumps may exceed 15,000. The great majority of these are the Anderson type standard Bush Pump. Most of these have 40 mm or 50 mm steel rising mains with 12 mm or 16 mm mild steel rods and are fitted with 50 mm or 75 mm brass cylinders. The new standard Bush Pump is also fitted with similar 'down the hole' components in most cases. Thus the majority of Bush Pumps use non-extractable valves and can be united in one category.

In 1986 an extractable valve system was designed for the Bush Pump by V & W Engineering, Harare, where the piston valve and the footvalve could be extracted through a 50 mm steel rising main.

In 1988, Blair Research Laboratory introduced the experimental use of heavy duty polyethylene pipe in combination with 12 mm drawn steel rods, in an attempt to demonstrate a less cumbersome technique for maintaining 'down the hole' components of the Bush Pump.

All three methods are described since the technique of installation varies from one to the next.

Method 1 *Fitting the Bush Pump with non-extractable 'down the hole' components*

This is the most widely used technique, and will be described for a situation where a standard Bush Pump ('B' Type) is equipped with a

50 mm steel rising main, 16 mm mild steel rods and a 75 mm cylinder and matching footvalve.

The main parts of the pump are:

1. Pump head complete with steel handle and hardwood block
2. Set of 3 metre length 50 mm galvanised iron pipes ('exacts')
3. Matching set of mild steel 16 mm rods
4. One 75 mm brass cylinder
5. One matching brass footvalve

Measurement of water depth and length of 'down the hole' components

In the case of a borehole, these are generally drilled much deeper than the final length of the pump; the exact length of the pump is not critical, provided it serves the users adequately. In the case of a well however, the pump footvalve should lie 0.5 metre above the bottom of the well, and thus an exact measurement of the well is required, and the 'down the hole' components should be built up to the correct length.

In any event, the required length of the pump should be known before installation, and the head should be provided with the appropriate number of 'pipes' and rods, taking into account that the cylinder and footvalve add between 0.5 m and 0.75 m to the total length of the column.

Measure the depth of the well or borehole.

When Bush Pumps are fitted to boreholes all the rising mains (pipes) and rods will be standard 3 metre lengths apart from the uppermost length which must be adjusted to accommodate the section of rod which passes beyond the rising main and enters the pump head. In many cases the uppermost rising main is cut short by a length of 400 mm, so that it is 2.6 metres long, thus giving the extra rod sufficient length to fit into the pump head. If standard 3 metre lengths of pipe are used throughout, an extra length of rod must be used for attachment to the pump head. This must be determined on site.

An 18 inch long 75 mm cylinder and matching footvalve add 700 mm of length to the rising main, and this must be taken into consideration when the pipe length is being estimated, especially for a well.

Inspection and assembly of components

All the components of the pump should be inspected and cleaned before they are assembled. The pipes should be laid out neatly on a stand to keep the threads clean, and each length of rod should be fitted within each pipe. Each pipe should be fitted with a socket and each rod with a connector.

Clean all parts of the pump.

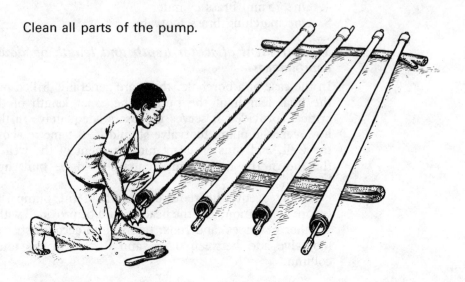

Stage 1

Clamp the pump head to the steel casing over the well or borehole. Remove the pump baseplate unit and put to one side. The pipes can now pass freely down the casing. If the pump is fitted with link arms and sleeve pipe, these must be removed and put to one side.

Place pump in borehole casing.

Clamp pump in position.

Stage 2

Inspect and clean the cylinder and footvalve. Fit together with a barrel nipple, using plumbers' paste. Take first rod and rising main, and attach the rod first, making sure the lock nut is tight. Now thread the first pipe to the head of the cylinder using plumbers' paste. The pipe and cylinder is now ready for lowering.

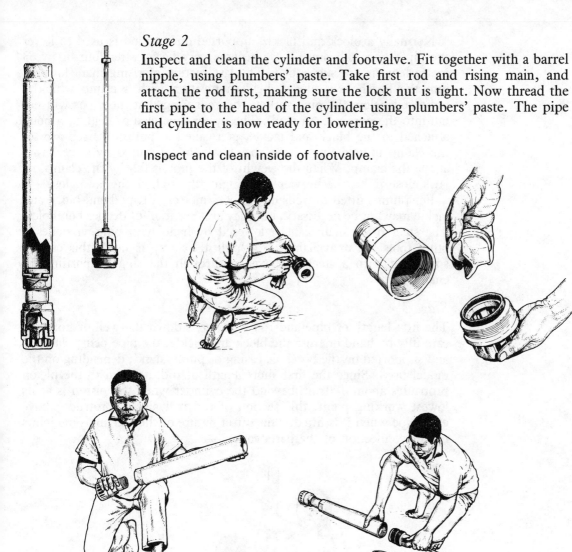

Inspect and clean inside of footvalve.

Attach footvalve to cylinder with barrel nipple.

Insert piston into cylinder.

Attach piston rod to first pump rod.

Attach first pipe to cylinder.

185

Normally a block and tackle supported on a tripod is used to lower the pipes and rods since these become very heavy after four or five 3 metre lengths are suspended. The last section of rising main lowered into the borehole is always held in position with a clamp which is tightened below the pipe socket, thus ensuring that the pipes will not fall into the borehole (or well). The upper pipe clamp, held by a hook attached to the block and tackle, is a called a 'pulltite'. Each arm of the clamp is held up by two chains which are united by a steel ring above the clamp. When the weight of the pipe is felt by the clamp, its jaws close tight, the heavier the weight, the tighter the jaws close.

For pumps fitted to wells down to 15 metres, a tripod and block and tackle may not be necessary, but they are essential for deeper boreholes. The steel rising main can be lowered by hand with pipe wrenches if other tools are unavailable. If a 'pulltite' clamp is used, this can be supported with a steel pipe passed through the ring supporting the chains.

Stage 3

The first length of pipe and rod is lowered down the well or borehole carefully by hand or from the block and tackle, the pipe being clamped and supported on the borehole casing or pump stand, depending on the model used. Since the first short length of rod attached to the piston protrudes about 100 mm beyond the cylinder when the piston is at its lowest working point, this section of rod will always protrude above each pipe when it is fitted. This offset nature of the rod and pipe joints makes connection of the parts easier.

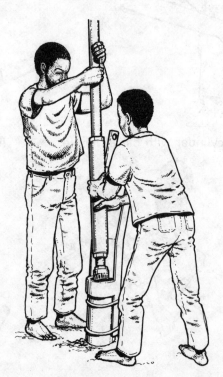

Lower cylinder and first pipe through pump head.

Stage 4
The remaining lengths of rising main and rod are now lowered section by section. The new pipe is raised with the rod inside it, and held up to enable the rods to be connected first, followed by the pipe. All rod joints should be tightened with a lock nut and all pipe joints should be packed with plumbers' paste. When both joints are secure, a new length is lowered down the hole.

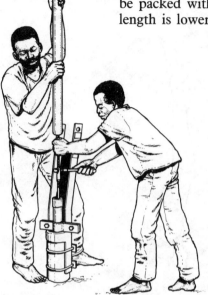

Holding second pipe up, join first and second rods.

Tighten first and second pipes together.

Hold up pipe whilst releasing and reattaching clamp.

Stage 5
The final length of rising main is clamped and threaded into the pump baseplate. The baseplate is now lowered and bolted to the pump stand.

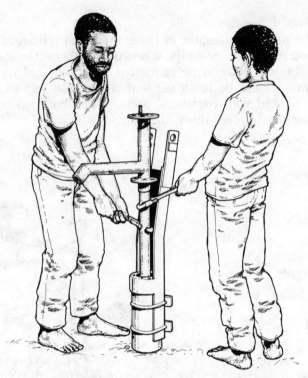

Attach uppermost pipe to pump baseplate assembly.

Bolt pump baseplate to pump stand.

Stage 6
The last pump rod should now be cut and threaded so that it lies level with the floating washers ('B' Type) or with the top of the guide pipe (earlier models). If it is too long it should be sawn off level, filed and the top 50 mm threaded with a 16 mm rod die. A piece of cloth should be packed around the pipe to stop cuttings from falling down the rising main.

Using U bracket, raise rod and hold it up with a vice grip or wrench. (Note: The rod must be cut and threaded to the correct length. When the rod is in its lowest position it should just show free of the floating washer housing as shown in previous illustrations.)

Once the rod is threaded, it is pulled up carefully as far as it will move and clamped with a vice grip. The technique of fitting the rods to the pump head will vary with each Bush Pump model. On older pumps the guide pipe, guide pipe washer, rubber buffer and sleeve pipe are now lowered over the rod. The rod is screwed through the threaded socket at the top of the sleeve pipe and held tight with a lock nut. The link arms are now attached to the sleeve pipe and the wooden block. The head block bolts are done up tightly and cotter pins added to the sleeve pipe assembly to retain the link arms.

On the 'B' Type Bush Pump, the rod is pulled up so that it passes through both upper and lower washers of the floating washer housing. The rubber buffer is now lowered over the rod and the front U bracket is threaded on to the upper end of the rod and secured with a lock nut. The U bracket is now secured to the wooden block with a large head bolt which is secured with a spring washer and single nut.

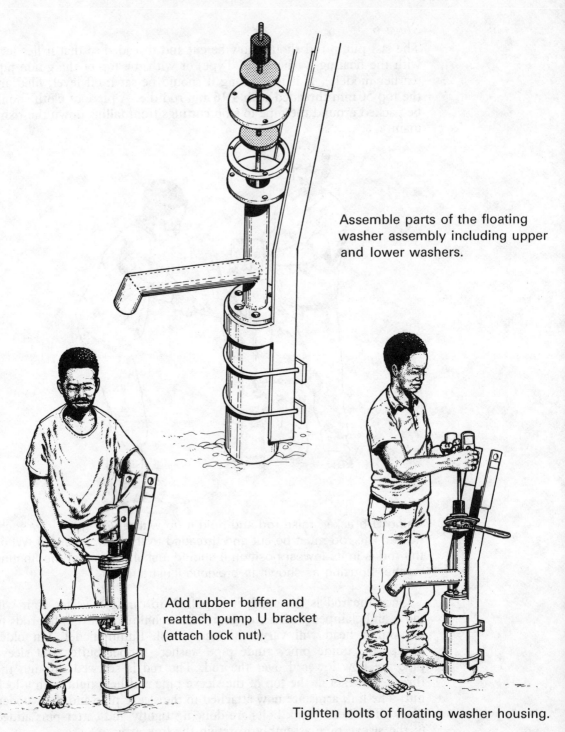

Assemble parts of the floating washer assembly including upper and lower washers.

Add rubber buffer and reattach pump U bracket (attach lock nut).

Tighten bolts of floating washer housing.

Stage 7
The head block bolts differ slightly depending on the model of Bush Pump used. In older pumps, a pivot tube is pushed through the wooden block and this is clamped between the link arms with a 24 mm bolt which is held by a nut and lock nut. In the newer Bush Pumps ('A' and 'B' Types) a solid head bolt is used (see diagram).

Fit wooden block and attach rear head bolt to pump.

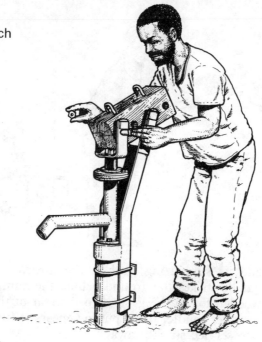

Stage 8
The handle is added to the hardwood block and clamped tight with the 12 mm U bolts.

Insert handle through two U bolts in wooden block.

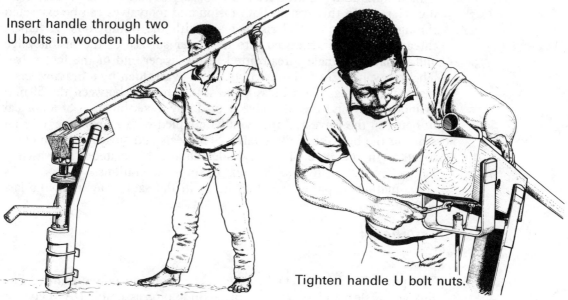

Tighten handle U bolt nuts.

Stage 9
The rear head bolt which supports the wooden block on the pump stand is checked for tightness. This should be already tightened to the pump head, but is should be inspected for smooth action and tightness. All other nuts and bolts should be checked for tightness.

Stage 10
The pump is now ready for testing. At first the water may be dirty but this will soon clear. The down stroke is buffered by the rubber buffer

and its supporting washer, the upstroke is buffered when the steel handle strikes the ground behind the pump. It is advisable to mount a wooden log in the ground to the rear of the pump to act as a stop for the steel handle. If the pump appears to function properly it can be put into service straight away.

Method 2

Fitting the Bush Pump with extractable valves

This technique was introduced in February 1986 by V & W Engineering, Harare. In this system both piston and footvalves can be extracted through the 50 mm steel rising main with the pump rods. A female thread has been introduced into the inner surface of the piston valve and a matching male thread fitted to the upper end of the foot valve. The foot valve uses a simple rubber disc overridden by a brass washer, making a seal against a brass valve seat. The seal between the 50 mm cylinder body and the footvalve is made by a combination of a conical fit between the valve and the cylinder coupled with 2 neoprene 'O' ring seals on the body of the footvalve. The seals and surfaces which come together in the base of the cylinder should be coated with vaseline before installation to ease removal in future operations.

Estimation of pump length is made in the same way as in the last section.

Installation

Stage 1

The pump head is attached to the steel casing and bolted tightly. The link arms, sleeve pipe, and baseplate/guide pipe assembly ('A' Type) or baseplate and water discharge assembly ('B' Type) are removed and put to one side.

Stage 2

The 50 mm brass cylinder, footvalve and piston valve are inspected, cleaned, and vaseline added to the lower valve surfaces to ease removal during future maintenance.

The cylinder is attached to the first length of rising main through a 50 mm socket with all joints being fitted together with plumbers' paste.

This system was designed by V & W Engineers, and enables both the piston valve and the footvalve to be extracted through a 50 mm steel rising main, the cylinder diameter is 50 mm.

In this system, the cylinder and 50 mm rising main are assembled and lowered first and the baseplate attached to the pump stand. The footvalve and piston are then assembled, attached to the pump rod, which is lowered down through the rising main. When being lowered the rubber buffer is not inserted into the system. After the footvalve has been pressed in to its seat, the two valves are separated by unscrewing them, and the rods and piston are raised and clamped. The rubber buffer and washer are inserted over the rod which is lowered into position.

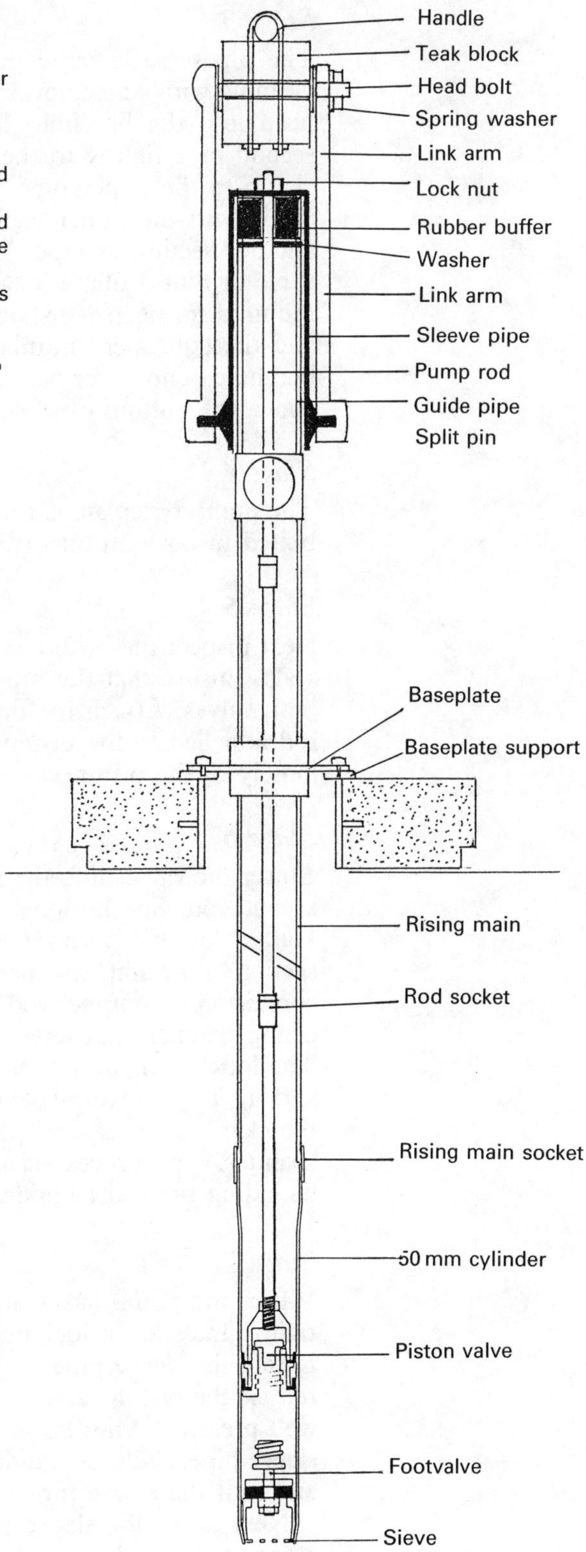

Tools

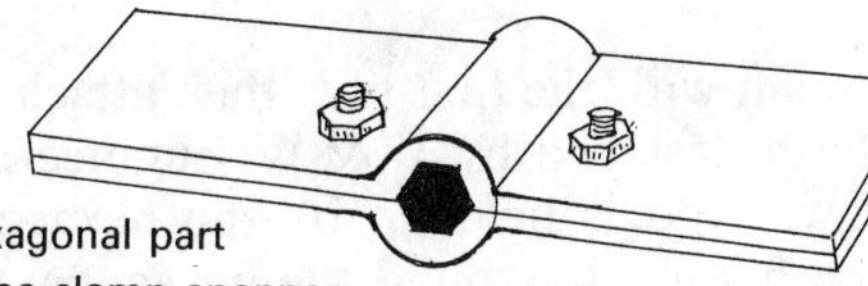

Hexagonal part of the clamp spanner

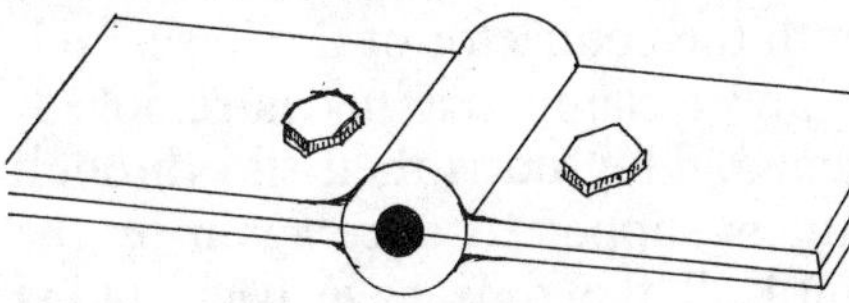

Clamp part of the clamp spanner

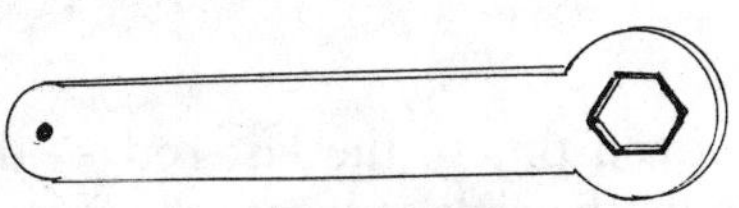

Large spanner

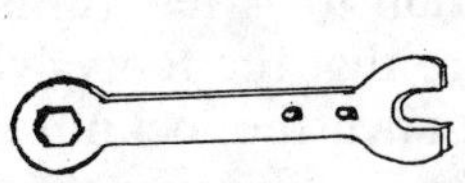

Small spanner with piston valve tool

Handle
Teak block
Head bolt
Spring washer
Link arm
Lock nut
Rubber buffer
Washer
Link arm
Sleeve pipe
Pump rod
Guide pipe
Split pin
Baseplate
Baseplate support
Rising main
Rod socket
Rising main socket
50 mm cylinder
Piston valve
Footvalve
Sieve

Extractable valve system.

193

Stage 3

The pipes are lowered into the well or borehole without rods. After cleaning with water, lower the first pipe and cylinder through the pump head into the borehole and support it with a base clamp. Attach a second pipe tightly to the first through the socket, and attach another clamp to the upper pipe. Loosen the lower clamp and lower the pipe again with the upper clamp. Attach the base clamp again and attach another section of pipe. This process can be performed by hand with three or four 3 metre lengths of 50 mm pipe, but a block and tackle is required for deeper installations. The number of lengths of rising main and the equivalent number of rods depend on the depth of the well/borehole. The lower part of the cylinder should lie at least half a metre above the bottom of a well or tubewell.

Stage 4

The pump baseplate is threaded on to the last length of pipe and this is bolted in position on the pump head.

Stage 5

Next inspect the piston and its neoprene seals and the footvalve and its seals. Ensure that the brass valves are free to move. Thoroughly clean both valves. Attach the first length of pump rod to the short length of rod attached to the piston. Tighten the lock nut. Attach the footvalve loosely to the piston valve and lower into the rising main.

Stage 6

Lower the valves into the rising main with the first rod, then attach the second rod. Special tools have been designed by V & W Engineering, Harare for this purpose. The lower rod is held up by the hexagonal section of a clamp spanner, which fits neatly around the hexagonal rod connector. The upper rod is clamped with the clamp section of another clamp spanner, and screwed tightly into the connector of the lower rod. The lock nut is now tightened. The lower clamp spanner is removed, and the upper clamp spanner is loosened. The rod is then slid through the clamp spanner until the connector is supported once again by the spanner. The process is repeated until all the rods have been added including the short uppermost rod.

Stage 7

Where the pump has a sleeve pipe attach this to the last rod — do up tightly and add the lock nut. Press down the rods to locate the footvalve. Rotate the sleeve pipe in an anti-clockwise direction for a few turns to release the piston valve from the footvalve. Ensure that the footvalve is well pressed home. Raise the rod, clamp and remove the locknut and sleeve pipe. Slide the guide pipe washer and rubber buffer over the rod and refit the sleeve pipe as before, doing up the lock nut tightly.

Next lower the sleeve pipe over the guide pipe. The rubber buffer acts as a shock absorber between the sleeve pipe and the guide pipe. In the extractable valve model it also acts as a spacer which separates the piston valve from the footvalve.

Where a floating washer system is used ('B' Type) the same sequence takes place, but the piston is unscrewed from the footvalve by attaching the rod to the U bracket and clamping with the lock nut. The U bracket is rotated anti-clockwise to separate piston and footvalves.

Stage 8
If the wooden block is not attached to the steel pump stand, this must now be fitted. Look at the diagram carefully before fitting the block. Note the heavy duty head bolt passing through the U bracket at the head of the pump stand and the bolt stabiliser that stops the bolt rotating. The heavy duty bolt is secured with a nut and spring washer, and tightened with a spanner.

Stage 9
Next the front end head assembly is fitted together. Once again study the diagram before fitting. One of the link arms is welded to the front block bolt. Fit both link arms over the sleeve pipe assembly at the same time as the front block bolt is passed through the wooden block. Attach the split pins. Once the link arms are fitted in position, the whole front head assembly is secured by adding the spring washer to the bolt and tightening this against the link arm.

Stage 10
The pump handle is now secured to the wooden block handle with U bolts provided. Tighten the U bolt nuts with their washers fitted.

Stage 11
The pump is now tested. It should pump freely.

Method 3 *Fitting the Bush Pump with polyethylene rising main and steel pump rods*

This is a relatively new technique in Zimbabwe and at the time of writing (April 1988) is still on trial. Heavy duty polyethylene pipe is used in the United States as a durable rising main for reciprocating water pumps. It is normally used in combination with fibreglass pump rods. However fibreglass pump rods are not available in Zimbabwe, and the combination of thick walled polypipe and drawn 12 mm steel (EN 8) rods, is being tried on an experimental basis.

It is possible to lower a single length of heavy duty polypipe, enclosing a single 'string' of rods into a well, within a few minutes. These can also be extracted without the need to disconnect any rods or pipe lengths. Examination of 'down the hole' components is made by pulling out the pipe in one piece. The cylinder is unscrewed and all working parts are examined directly. To date the system has been tried to depths of only 20 metres.

Special steel thread to polypipe adaptors with a large supporting area have been designed and the 12 mm rods joined together with normal connectors, although the threaded joints are bonded with epoxy resin cement. The steel to pipe adaptors are attached to the baseplate of the pump head and the pump cylinder the exact length of the pipe being cut to suit the rods which are cut to suit the well.

Installation procedure

The following parts are required:

1. Pump head
2. Suitable 3 m lengths of 12 mm EN 8 drawn steel rods with connectors
3. Suitable length of 40 mm or 50 mm heavy duty black polyethylene pipe
4. Two polypipe to steel thread adaptors
5. Brass cylinder and footvalve assembly (50 mm or 75 mm)
6. Two 12 mm to 16 mm rod adaptors
7. Two 40 mm to 50 mm GI reducing sockets (if 40 mm polypipe used)
8. Quick setting epoxy adhesive for rod joints
9. Four pipe jubilee clips

Stage 1

Test depth of well, tubewell or borehole. Assemble pump rods of suitable length and suitable length of polypipe.

Stage 2

Fit pump head on to casing at well head, remove link arms, sleeve pipe, baseplate/guide pipe ('A' Type), or baseplate and water discharge assembly ('B' Type) and put to one side.

Stage 3

Fit together pump rods. All threaded joints should be epoxy bonded together (fast setting Trinepon 6 in Zimbabwe). Since the pump head, and often the piston valve, is designed for 16 mm rods, 12 mm to 16 mm adaptors are necessary to convert from one rod size to another. The piston valve can be fitted to a half metre length of 16 mm rod attached to a 16 mm/12 mm adaptor. The sleeve pipe or U bracket of the pump head should be fitted to a one metre length of 16 mm rod, and then adapted to 12 mm. The exact length required should be judged on site, and the rod cut and threaded with a rod die as required.

Stage 4

Attach the upper end of the rod to the pump head. Tighten the lock nut. Locate the rod in the lowest position.

Lay the string of rods along the ground. Lay the piston and cylinder assembly next to the rods in the appropriate position, so that the piston valve is at the bottom of its stroke. Lay the pipe to thread adaptors next to the pump baseplate and the cylinder. If the cylinder is a standard 50 mm type, this will have a 40 mm thread attachment suitable for the 40 mm pipe adaptor. 75 mm cylinders have 50 mm threads — thus a 40 mm to 50 mm reducing bush is required at this joint if a 40 mm pipe is used. Similarly the baseplate of the standard Bush Pump is designed to take a 50 mm thread. If a 40 mm polypipe is used another 40/50 mm GI socket will be required.

Stage 5
With the pump head components, rods and cylinder laying in position on the ground, it will be possible to determine where to cut the polypipe. The polypipe is cut and attached to the adaptors. When the pipe has been lying in the hot sun, it becomes sufficiently flexible to attach to the adaptor without further heating. Add two jubilee clips to each joint and do up tightly.

Stage 6
Lift the entire length of the rods including the baseplate assembly (several men will be required to do this, since the rods are flexible), and take them to the head of the polypipe. Pass the entire length of the rods through the polypipe and screw up the adaptor at the upper end into the baseplate.

Stage 7
Unscrew the hexagonal cylinder end cap from the cylinder and attach this to the lower adaptor. Fit the lower end of the rod to the piston and tighten securely. Attach and tighten the body of the cylinder to the cylinder end cap. Attach the footvalve if this is not already attached. The rising main and rods are now ready to lower into the well or borehole.

Stage 8
This is a very special part of the procedure, and must be performed with care since the bend in the pipe/rod assembly must be kept to the minimum. This part of the process can be assisted by means of three poles or rods fitted at the upper end with angle iron large enough to support the pipe well above ground level.

Several men are required, and the pipe is lifted up along its length. The pipe is raised up on the supporting rods, and by hand so that it forms a large loop. The cylinder is fed through the pump head into the well, followed by the rest of the pipe. When the entire pipe has been lowered the base plate is bolted to the pump head and the rest of the pump assembled as in the previous section.

A certain amount of practice may be required to develop a good technique, but once mastered this method greatly eases the problem of maintaining 'down the hole' components.

Maintaining the Bush Pump

The success of the Bush Pump in providing water over long periods of time is dependent on the maintenance system which supports it. If the maintenance system, backed up by adequate staff, tools, stores and transport breaks down, so does the rural water supply.

The Bush Pump has been maintained by the District Development Fund (DDF) of the Ministry of Local Government, Rural and Urban Development for many years in Zimbabwe. The DDF operate at district level where pump fitting gangs, under the supervision of a District Field Officer are equipped with special tools to maintain the Bush

This technique is currently on trial. It enables the entire rising main to be extracted with the rods in one piece and thus makes inspection of the cylinder and valves far easier compared with techniques that use steel rising mains.

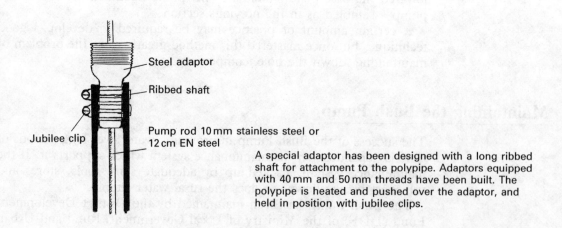

A special adaptor has been designed with a long ribbed shaft for attachment to the polypipe. Adaptors equipped with 40 mm and 50 mm threads have been built. The polypipe is heated and pushed over the adaptor, and held in position with jubilee clips.

Bush Pumps with flexible rising mains.

Pump and other water installations. More recently Pump Minders have been employed at Ward level in some areas to maintain handpumps, but this system of maintenance is still being evaluated. Many districts have more than 500 handpumps in operation and this places a big responsibility on the DDF and its staff.

However, villagers and pump caretakers have an important role to play in the maintenance of the Bush Pump by ensuring that all the nuts and bolts on the pump head are tight. This simple operation is very important, and can best be carried out by those on the spot. For this reason it is wise policy for suitable spanners to be left on site with a pump caretaker or chairman of the village water committee. Formerly, the pump head was greased by pump caretakers, but this is now thought to be unnecessary. The village should also be responsible for keeping the apron and run-off channel clean.

Currently the number of Bush Pumps operating in Zimbabwe is estimated at 15,000. The cost of maintaining these pumps is estimated at over Z$2 million per year (At time of writing this is equivalent to 1 million US dollars per year.) It is also estimated that 45% of this cost is used up in transport, 37% in labour and the remaining 18% in spare parts. On average each pump costs at least Z$150 (equivalent to 75 US dollars) per year to maintain. These high figures reflect the simple fact that each Bush Pump must be serviced by well equipped gangs who, because of the remote positions of many pumps, must travel great distances to undertake repairs. Under these conditions the gangs can only respond to reports of breakdown. In practice they are not in a position to perform preventative maintenance on each pump on a routine basis. Simple preventative maintenance, such as nut and bolt tightening on the head, must be carried out on the spot by the village water committee.

Most visits are made to Bush Pumps to replace the piston seals, which can wear away in as little as six months in heavily used installations. However, in normal settings the seals require replacing every two years. Failures of the rods and rising main are not uncommon, especially in ageing pumps. The inspection of leaky valves, especially poor quality footvalves also accounts for many visits being made to Bush Pumps. However, many pumps break down and require reassembling simply because the parts were not fitted together properly in the first place. It is important that Bush Pumps are repaired by those who are well trained, and have a knowledge of how the pump works and where the potential problems of breakdown may occur.

Maintenance of the head assembly

This is a robust unit, its main requirement being the tightening of all nuts and bolts, a procedure that can be carried out at village level with limited training and tools.

Main head bolts

These must be kept tight. On older pumps the pivot tubes must be held secure within the link arms or pump stand. If these come loose, the tube may tighten within the block and will rotate around the bolt, leading to wear. If the bolt is loose it will move about, either in the link

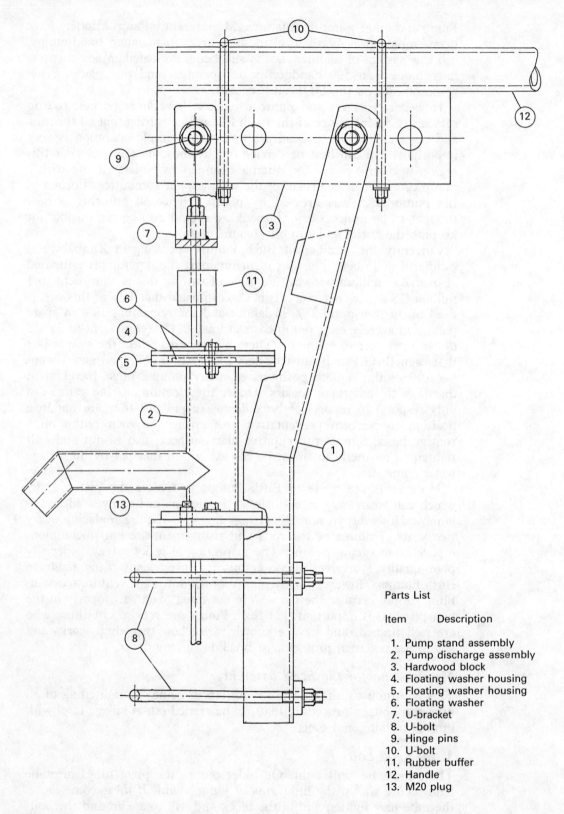

Parts List

Item	Description
1.	Pump stand assembly
2.	Pump discharge assembly
3.	Hardwood block
4.	Floating washer housing
5.	Floating washer housing
6.	Floating washer
7.	U-bracket
8.	U-bolt
9.	Hinge pins
10.	U-bolt
11.	Rubber buffer
12.	Handle
13.	M20 plug

'B' type Bush Pump.

arms or the U bracket, leading to wear and the necessity for eventual replacement. In more recent models, wood to steel bearing surfaces are guaranteed, since the main bolt cannot turn. However for long life, the bolt must be kept tight against the spring washer.

It is wise to remove the head bolt from time to time, apply a layer of grease to the shaft and run this through the hole in the block a few times, then replace and tighten the bolt.

Sleeve and guide pipes

These represent a metal to metal surface, and wear is inevitable on these parts in time. They must be checked and when badly worn — replaced. This may take several years.

Periodically, the rubber buffer and guide pipe washer should be checked. This is normally done when the piston seal is replaced, and the pump taken apart. These parts can wear badly, but are not visible from outside. Spare rubbers and washers should be available. The lock nut on the end of the rod must also be kept tight.

Floating washers

Two floating washers are fitted to the latest 'B' Type Bush Pump. In normal use these will require replacing every two years. The rate of wear depends on the amount of usage. Each pump is sold with two spare washers.

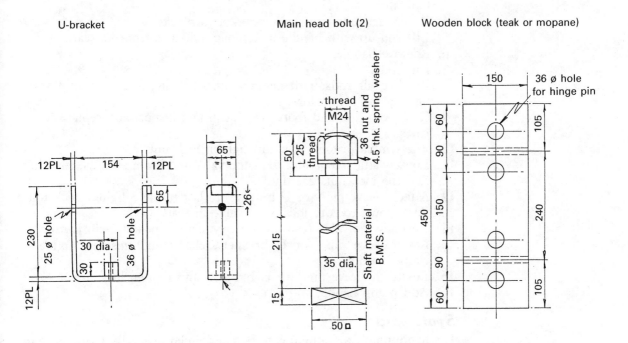

'B' type Bush Pump.

Replacement of non-extractable seals and inspection of 'down the hole' components.

1. The pump head should be dismantled.
2. Lift the first length of rising main and rods by attaching clamps and lifting either by hand or with a block and tackle.
3. When the first socket is reached, base-clamp the rising main just under the socket, and unscrew the first length of pipe with pipe wrenches. Raise this pipe to expose the rod and unscrew the rod connector. Remove the disconnected parts and repeat by clamping and raising the second pipe. Base-clamp again under the socket and repeat the procedure.
4. Remove all pipes and rods to expose the cylinder and valves. These should be dismantled, cleaned and checked for wear or breakage. Worn parts including seals should be replaced by new parts. All parts should be cleaned, reassembled carefully and tightened.
5. Reassemble the rods and rising mains as described earlier, making sure that all threads are packed with plumbers' paste and all rod connectors secured with a lock nut.
6. Reassemble the pump head. Ensure all nuts and bolts are tight.

Replacement of extractable seals and inspecting 'down the hole' components

1. Dismantle pump head.
2. Remove the rubber buffer. (This lowers the piston valve so that it can be attached to the footvalve).
3. Turn sleeve pipe (or U bracket) clockwise several times to pick up footvalve.
4. Reconnect link arms or U bracket to block.
5. Lift rod up with handle to remove footvalve from its seat.
6. Remove handle.
7. Lift rods.
8. Remove all rods with clamps, untightening lock nut first then unscrewing connectors.
9. Remove piston and footvalve. Clean and inspect and replace worn parts.
10. Reassemble piston valve and tighten lock nut.
11. Loosely attach footvalve to piston valve and lower all rods, ensuring that the lock nuts are tight.
12. Attach sleeve pipe (or U bracket) to last rod and tighten lock nut. Push down rods to locate footvalve in seat.
13. Unscrew footvalve from piston by turning rods anti-clockwise.
14. Raise sleeve pipe (or U bracket), clamp and unscrew and add rubber buffer.
15. Re-attach sleeve pipe (or U bracket) and tighten lock nut.
16. Add pump handle and test.

Spare parts

If a handpump programme is to be maintained properly, spare parts to fit them must always be available.

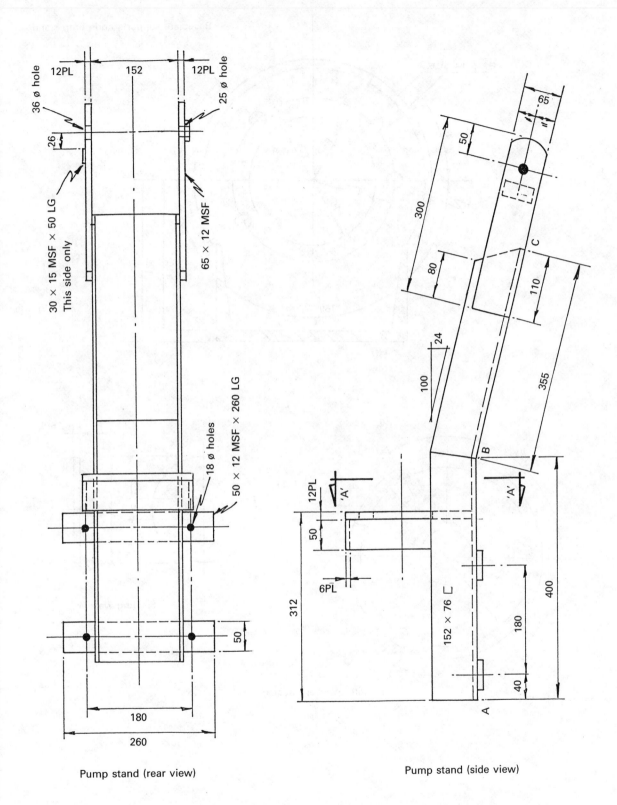

Detail of 'B' type Bush Pump.

Baseplate supports and pump stand

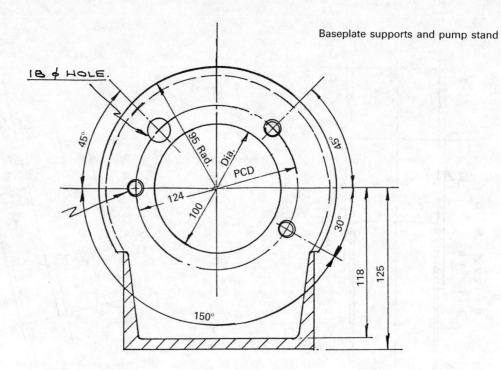

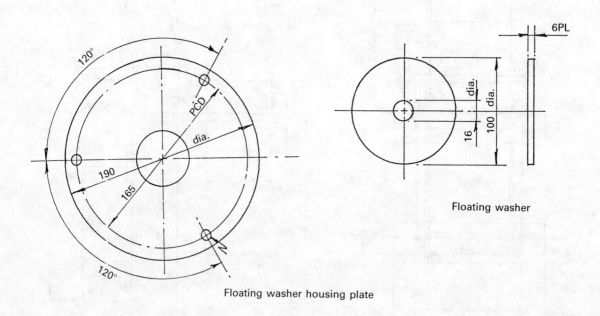

Floating washer housing plate

Floating washer

Detail of 'B' type Bush Pump.

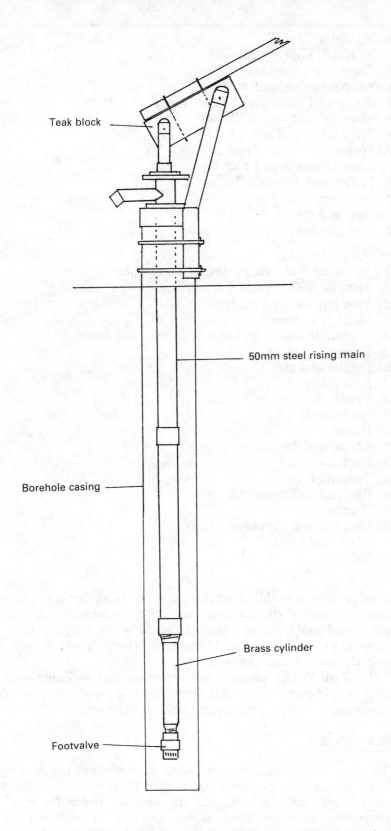

'B' type Bush Pump with 65 mm extractable piston

These include:
1. Rods and rising mains
2. Piston seals (leathers)
3. Rubber buffers and washers
4. Sleeve and guide pipes (older pumps)
5. Floating washers ('B' Type)
6. Pivot tubes and block bolts (on older models)
7. Spring washers (on new models)
8. Brass cylinders and footvalves
9. 12 mm and 16 mm U bolts
10. Link arms
11. Pipe sockets
12. Rod connectors

Tools

The following tools should always be available.
1. 610 mm, 450 mm and 350 mm wrench spanners
2. Vice grip
3. Shifting spanner
4. 12 mm, 16 mm and 24 mm open ended spanners
5. Pliers
6. Pulltite clamp
7. Base clamp
8. Chisel
9. Wire brush
10. Hammer
11. 12 mm and 16 mm rod dies
12. Hacksaw
13. Tape measure
14. Flat and half round file
15. Plumb line
16. Oil can (and plumbers' paste)
17. Block and tackle
18. Tripod
19. Rags

It can be seen from this description that the Bush Pump requires many tools and spare parts and much expertise to maintain. This contrasts very considerably with the Bucket Pump, where very few tools, spare parts and minimal expertise are required. The Blair and Nsimbi Pumps lie in an intermediate position.

The Bush Pump remains vital to the success of Zimbabwe's rural water supply programme, and every possible step should be taken to ensure that it receives the maintenance it deserves.

Bush Pump — Research

The great success of the Bush Pump in Zimbabwe can be attributed to several factors. Firstly it has always been strongly supported by Government, and has remained the national pump for over half a century. Secondly it uses very sound principles. A hardwood block acts as a lever and bearing surface and this has been supremely successful.

Thirdly, most of the working parts of the head are exposed and can be attended to easily, and can even be substituted by non-standard parts. This means that the pump head can be made to operate, in some way, even under extreme conditions of wear. Fourthly the pump head is rugged, and can withstand extremely heavy working conditions.

The Bush Pump, as a whole, is not without its faults however. The sleeve pipe concept of attaching the wooden block to the pump rod is subject to wear, especially when the pump head is poorly aligned, and it could be argued that this arrangement is unnecessarily complicated. The pivot tube, used in older pumps, had a tendency to jam within the teak block, resulting in a loss of the elegant wood/steel bearing concept. In older pumps the correct alignment of the pump head and rising main could never be guaranteed. The use of steel rising mains with a smaller diameter than the pump cylinder, necessitates extraction of the entire rising main and rods, to attend to 'down the hole' servicing. Replacement of piston seals, for instance, can only be carried out once the rising main and rods have been hauled out. This is a procedure which requires block and tackle and shearlegs, especially for deeper pumps. Thus the replacement of parts often necessitates the use of heavy transport and skilled personnel.

As the number of pumps increases, currently at the rate of about 2000 per year, there is an ever growing need to look at the efficiency and cost of maintenance. Several aspects of Bush Pump technology are currently being investigated for this reason.

Bush Pumps with extractable valves

The relatively high costs of maintaining Bush Pumps is mainly due to high transport costs in conveying trained staff and heavy equipment (shearlegs and block and tackle) to the remote sites where Bush Pumps are often located. Whilst the pump head can be serviced adequately by keeping bolts tight, a task that can be performed by the local pump committee, repairs and maintenance of 'down the hole' components is not so easy, and must be carried out by trained pump minders or pump fitters.

Various attempts have been made to make the servicing of 'down the hole' components an easier task. One attempt has been the adoption of fully extractable valves, which can be extracted through the rising main. One extractable valve system in use was designed by V & W Engineering in 1985, and is illustrated earlier in the text.

The system is based on a 50 mm cylinder with strengthened piston valve assembly carrying two neoprene seals. The lower end of the piston has a wide achme thread which can be lowered and screwed on to a male thread which forms the upper part of the footvalve which is press fitted in the lower end of the cylinder. The upper end of the 50 mm cylinder is expanded and fitted with a standard 50 mm thread for attachment through a 50 mm socket to the 50 mm GI rising main. Both piston valve and footvalve can be extracted through the rising mains. Both 12 mm (EN 8) and 16 mm (Mild Steel) rods are used with this system. It is important that the string of rods does not separate if this system is to have its full value. This extractable valve system is still being evaluated.

Table 7. Frequency of main Bush Pump breakdown/repairs. Data of DDF maintenance teams 1987–1988.

TYPE OF REPAIR	NO. OF REPAIRS	PERCENTAGE TOTAL REPAIRS
1. Leather seals	1055	24%
2. Rising mains	635	15%
3. Rods	486	11%
4. M 24 bolts	250	6%
5. Footvalve	238	6%
6. Cylinder	164	4%
7. Guide pipe	154	4%
8. Tee piece	152	4%
9. Bracket arms	139	3%
10. Sleeve pipe	120	3%
11. Base plate	110	3%
12. M 12 bolts	109	3%
13. Pump stand	92	2%
14. M 16 bolts	72	2%
15. Wooden block	61	1%
OTHERS	440	9%
TOTAL	4277	100% (64% below ground) (36% above ground)

How the 'B' type Bush Pump with extractable 65 mm piston attempts to reduce these problems.

PROBLEM	ATTEMPTED SOLUTION
1. Leather seals	More easily extracted
2. Rising mains	Larger size — less frequent extraction
3. Rods	Longer connectors anticipated
4. M 24 bolts	New solid bolts used in 'B' type
5. Footvalve	Reliable heavy duty type chosen
6. Cylinder	External threads reduce cracking
7. Guide pipe	Not used
8. Tee piece	Not used
9. Bracket arms	Not used
10. Sleeve pipe	Not used
11. Base plate	Welded to pump stand
12. M 12 bolts	No change
13. Pump stand	Shorter/higher quality
14. M 16 bolts	No change
15. Block	Higher quality

Analysis of these figures show how attempts are being made to reduce the number of breakdowns and simplify maintenance procedures by modifying the design of the pump head and 'down the hole' components.

The advantage of this system is that the 'down the hole' working parts can be extracted relatively easily by hand. Also since the steel rising main is left undisturbed, the thread life is extended, which prolongs the life of the pipe itself.

50 mm cylinders of this type yield between 15 and 20 litres of water per minute, about half the yield of a 75 mm cylinder. This is not a big disadvantage where the pump is used by relatively small numbers of people, but is inadequate where a single pump serves a large community and is required for cattle watering or irrigation. Here the 75 mm cylinder may be essential. Parts on the 50 mm cylinder appear to wear out faster than on the 75 mm unit, probably because they are smaller. The 75 mm cylinder is preferred by users since it delivers more water, although it is harder to pump. Those involved with maintenance must consider the costs of keeping the pump operating, as well as delivery rates. 50 mm cylinders are essential when placed on very deep boreholes, since the larger type is too hard to pump. Currently tests are being carried out on a 65 mm diameter extractable version, where water output is improved and the parts are larger and thus more durable. This system can be used with 65 mm GI rising main, flexible polyethylene pipe and PVC pipe.

Bush Pumps with flexible rising mains

For many years Bush Pumps have been fitted with 50 mm galvanized iron rising mains equipped with 16 mm mild steel rods. More recently 40 mm GI rising mains have been introduced in some areas to reduce cost. However, although the smaller pipe is cheaper and lighter, inspection and replacement of 'down the hole' components is still not easy.

Alternative means are being explored to make inspection and replacement of 'down the hole' components easier. One method, described above, enables the valves to be taken out through the rising main. This method is still being perfected. A second method employs the use of thick walled, heavy duty polyethylene pipe in combination with flexible 12 mm EN8 steel or stainless steel rods. With care, the rod can be extracted within the pipe in one piece, together with the cylinder. Development of this process is in its infancy in Zimbabwe, but current results are encouraging. A variety of techniques using polypipe with diameters ranging from 40 mm to 92 mm (ID) with different rod types are being investigated. One technique is described under installation and maintenance of Bush Pumps.

Supplying water by gravity

The most elegant method of supplying water is not by pumping or even lifting it from underground, but by making it flow under the influence of gravity to where people live. Gravity has been used for thousands of years to take water from the well-watered mountainous areas of Iran into the dry but fertile plains through a remarkable series of underground channels called qanats. Although many of these were built 3000 years ago, some still serve today as a means of supplying water. Some are nearly 30 km long, others much less. Some deliver 270 litres per second, others 1 litre per second, but they all have one thing in common — they are reliable and long lasting.

Bearing in mind that the most important thing about a rural water supply is its reliability, it is surprising that gravity schemes have not played a more important role in Zimbabwe. The most probable reason for this is that most of the water reservoirs found in the Communal Lands do lie underground, and that sites where water is found above the level where people live, such as springs, are not common in most areas. The harvesting of water from rainwater catchment areas, like roofs, is also less common than in countries like Kenya, because Zimbabwe has only one rainy season and a long dry spell in between, when the management of stored water becomes more difficult. By comparison rainwater catchment tanks are common in Kenya, which has two rainy seasons each year.

However, there are many examples in the rural areas of Zimbabwe that show that water can be gravitated successfully to supply domestic water, and even sufficient water for small scale irrigation. In Manicaland, for instance, many hundreds of small springs have been protected in the hills, the water collected being piped to communal standposts within a village. Also there are examples of where small mountain streams have been dammed and the water collected in this way has been piped to supply overhead irrigation for crops lower down. There are also many examples of where rainwater has been collected on the roofs of dwellings and stored to provide domestic water.

Successful gravity schemes are not confined to Manicaland. Over the years large numbers of springs have been protected in many other provinces and still provide water, reliably, to this day.

This section of the book describes the various techniques of providing water by gravity. It includes the protection of naturally occurring springs, the construction of the siphon well and gravity well, and the harvesting and storage of rainwater. The design of a piped water point is also described, as a tap outlet is often provided at the user end of the gravity scheme. Methods of disposing of waste water produced at the water point are also described. The full health benefits of an improved water supply cannot be achieved if the surroundings of the supply are unhygienic.

Wherever possible the gravity system of supplying water should be used, as this is the most reliable means of providing water in the long term.

The protection of naturally occurring springs

Springs are found mainly in mountainous or hilly areas and occur in specific sites where the groundwater is not permitted to remain underground and comes to the surface. Spring water is usually fed from water-bearing ground formations composed of sand or gravel or from fissured rock which overlays a base of clay or impervious rock. The water runs through the layer until the aquifer meets the surface or is forced up under pressure from above. The diagrams below represent some situations in which springs are formed and occur naturally.

Spring water may emerge either directly in the open as in a gravity-

The natural occurrence of springs.

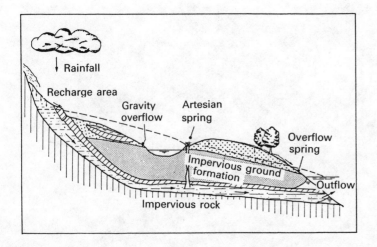

Note: The presence of a recharge area above, an impervious layer below and zones in which water finds its way to the surface, either by gravity or by force up an artesian spring.

Here the recharge area directs water through the lower aquifer and up again into an artesian spring.

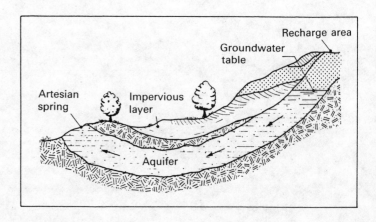

Illustrations taken from
Small Community Water Supplies
WHO-IRC Technical Paper No. 18
1981.

fed spring on a mountain side or may be less apparent as in the case of an overflow trickling almost out of sight into a stream or lake. Spring water can easily be tapped and is indeed one of the most ancient methods of collecting naturally occurring purified water. Springs can often be identified by patches of particularly lush vegetation on the hillside. They are of course very well known by traditional folk and are often associated with folklore in rural areas. Springs have often been the source of water for many people over the centuries, and great care is required to ensure that the process of protecting a spring does not result in its drying up or finding an alternative course. The family owners of many springs in Zimbabwe may prefer that the spring is left untouched, and this preference must be adhered to.

Spring water is generally pure and may only be contaminated at the spot where it emerges if there is constant human or animal contact. The aim of spring protection is to collect the purified water and pipe it to a spot lower down, which will become the site of human and animal contact. This ensures that the site of the spring itself, which is covered over at the surface, is subject to very little contamination.

The flow of water from a spring may be through openings of various shapes. Filtration springs occur where the water percolates from many small openings in porous ground. Fracture springs occur in areas where water issues from joints or fractures or otherwise solid rock. Tubular springs occur where the outflow opening is rounded. Here the tubular capillaries open into the spring at the 'eye', a familiar word in spring technology. The art of making a good spring is to maintain the integrity of the 'eyes' so that the yield from this source is maintained.

It is necessary also to distinguish between gravity springs and artesian springs. Gravity springs occur in so-called unconfined aquifers where the ground surface dips below the water table. Such depressions will be filled with water. Gravity depression springs usually have a small yield and they are influenced by seasonal variation in yield. A larger and less variable yield from gravity springs occurs where an outcrop of clay or other impervious material prevents the downward flow of water. Water is discharged as an overflow in these cases at the surface. Artesian depression springs are those where water is forced to the surface under pressure. All artesian groundwater is under pressure because the source of the water is at a higher point and the aquifer is impervious from above and below. The artesian spring (or well) occurs where the impervious upper layer is broken and the water can rise through this break.

The protection of springs

All naturally occurring springs are exposed to the surface and if they are much used can invite areas of contamination around them by human and animal contact. The aim of protecting a spring is to tap off the water carefully from the 'eyes' or infiltration area, allow it to build up in an enclosed reservoir which is capped, and to arrange for the water to overflow from the reservoir to a lower point, away from the original site of the spring so that the natural spring is undisturbed.

The exact procedure obviously depends on the type of spring. A very large number (many hundreds) of naturally occurring springs have been protected and serve as gravity-fed water supplies in the Manicaland Province of Zimbabwe and to a lesser extent elsewhere. At the present time the art of protecting springs is not well known and deserves much more attention. The most successful and long lasting of all man-made water supplies are springs or at least gravity-fed water supplies. The unique simplicity of the gravity water supply, of which the spring is the best example, is its greatest strength and deserves far more attention.

Tapping and protecting gravity springs

Gravity overflow springs can be tapped with drains built of stonework, loose brickwork or concrete rings set in a gravel pack which adjoins the water-bearing surface. Here it is necessary to dig carefully into this area of the spring and to note where the 'eyes' and other water bearing zones are. It is necessary to dig into the hillside so that a sufficient depth of the aquifer is tapped even when the groundwater is low. Water may percolate into the spring chamber from the side walls or from below, or from a combination of both. Naturally occurring capillaries and their exposed eyes can yield considerable quantities of water — and as the excavation continues a series of small fountains of water may 'spring' from these eyes. During the excavation it is important to keep all these 'eyes' open. If an 'eye' is blocked, some of the water in the aquifer may be diverted elsewhere.

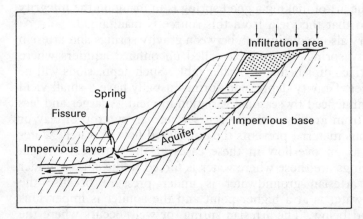

Here the recharge area feeds water into the aquifer which becomes narrowed below, water passes under pressure through any crack or fissure in the rock.

Artesian depression spring.

Here the aquifer is directly exposed to the surface on a slope — normally in hilly country.

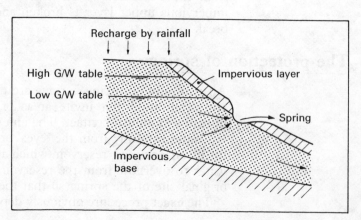

The construction of the reservoir — often referred to as a 'spring box' — and its surrounding gravel pack will depend to a certain extent on where the water is coming from. If most of the 'eyes' are situated on the inner wall of the excavation, it will be necessary to pay particular attention to laying a good gravel pack in that area and plugging other areas between the 'box' and the outer wall with clay, so the water that builds up in the reservoir cannot overflow or drain through the soil and be lost. If the 'eyes' or infiltration area seem to lie at the base of the excavation, it will be necessary to set the gravel pack on the base of the excavation, thus allowing water to pass through it upwards. The reservoir is constructed on top of the gravel pack. In this case the space between the reservoir and the side walls of the excavation can be packed with clay all round to prevent the loss of water from the reservoir by seepage.

In all cases two pipes are cast into the reservoir — one at a lower point which is the main discharge pipe and one at a higher point, which is an overflow pipe. In some cases, a third pipe can be fitted below the discharge pipe which acts as a scour pipe, which can be used to clean out the spring from time to time.

Diagrams of typical arrangements for protecting a spring are shown on the following pages. Once the stone wall or brick wall or concrete rings have been laid in position and surrounded by the necessary gravel pack, which usually consists of coarse river sand or gravel, a concrete lid is prepared for fitting on the top of the 'spring box'. Normally steel piping is used to convey water through the wall of the 'box', and heavy duty polypipe or PVC from the 'box' to the water point. All exposed

An example of a spring water storage chamber or 'spring box'.

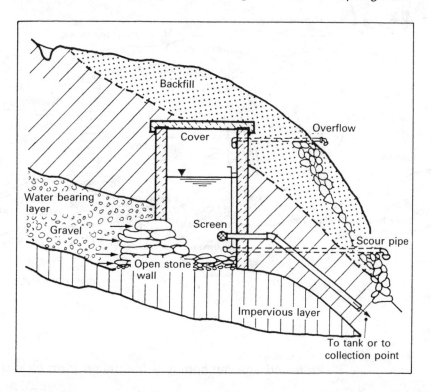

Protected spring infiltration from back wall.

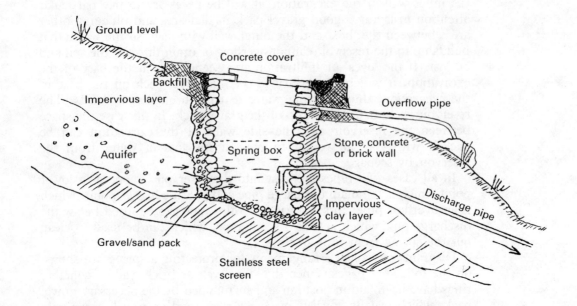

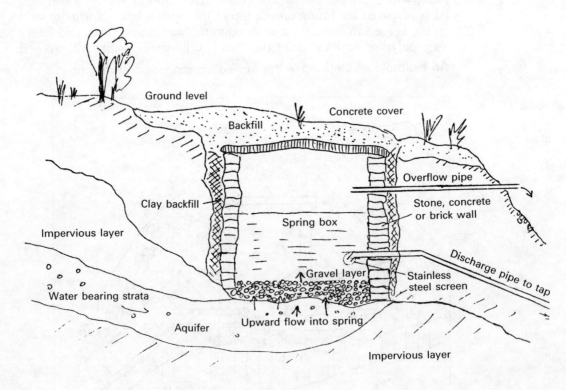

Protected artesian spring — infiltration from base of reservoir.

piping should be made of steel. The piping is led downhill to the most convenient spot away from the site of the natural spring. This may be only a few metres away, but in some cases it may be several hundreds of metres away. Very often a tap is fitted to the end of the main discharge pipe so that when the spring water is not being used the reservoir can slowly fill up. Where the spring yields large volumes of water, the water may be allowed to run continuously from the end of the discharge pipe. A tap will not be required in this case, and this reduces further the maintenance requirement of the facility.

When a spring is properly protected and adequately dug, a reliable delivery of water can almost be guaranteed. There are no valves, levers or seals, and no other moving parts. The water moves through the system as a result of gravity alone and for this reason should provide long and trouble-free service for many years.

Manicaland spring siphonage system

Stage 1

Locate the site of the spring on the hillside. Dig down into the site of the spring to a depth of 3 or 4 metres exposing the 'eyes' of the spring.

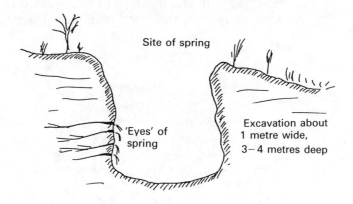

Stage 2

After the spring chamber has been dug and cleared lay down a layer of water-tight clay (Dongo) on the bottom of the excavation where no 'eyes' are present.

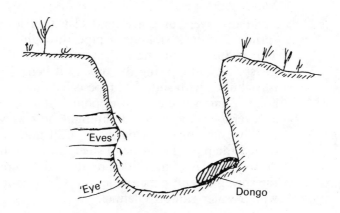

Stage 3

Build up the mouth of the spring with granite stones, bricks or concrete rings. Build it up a few metres above the base layer.

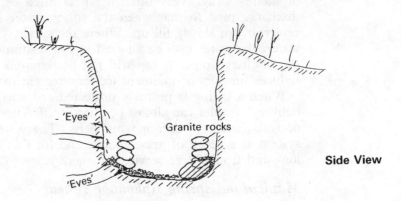

Stage 4

Pack the space between the stone, brick or concrete wall and the 'eyes' of the spring with granite chips or well washed coarse river sand. Pack the areas away from the 'eyes' with clay (Dongo).

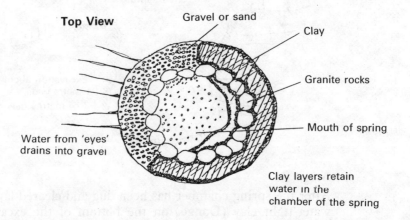

Stage 5

Build up layers of gravel and clay to a level above the 'eyes' of the spring. Lay in a siphonage pipe through the clay wall of the spring. This is done by cutting a channel through the soil on the lower side of the spring and burying the pipe in it beneath the ground. The channel is backfilled with soil. The pipe is led down to a draw-off point fitted to a tap or gate valve at a lower point. At a higher point in the spring box an overflow pipe is fitted through the clay wall. Both discharge pipe and overflow pipe can be made of 20 mm pipe. Where strength is required the pipe should be of steel, underground it can be made of PVC or polypipe. A 75 mm thick reinforced concrete slab is made to fit over the spring box. A concrete apron is built around the head of the spring with a water run-off channel.

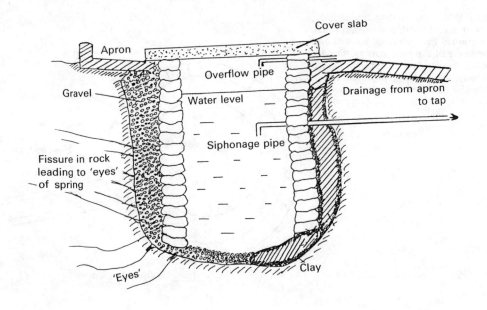

Stage 6
The siphonage pipe is led down to the water draw off point. This is an area dug into the ground and lined with rocks or bricks. Steps are made down to the tap.

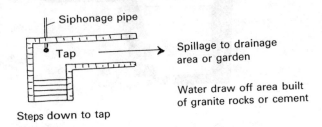

Stage 7
For an overhead view of protected spring see next page.

Note: Where an existing traditional spring is protected, permission must be granted by traditional leaders and members of the local community, or family owner. Any ceremonial tokens already laid in the spring must be carefully removed and placed back in the spring once the upgrading process is completed.

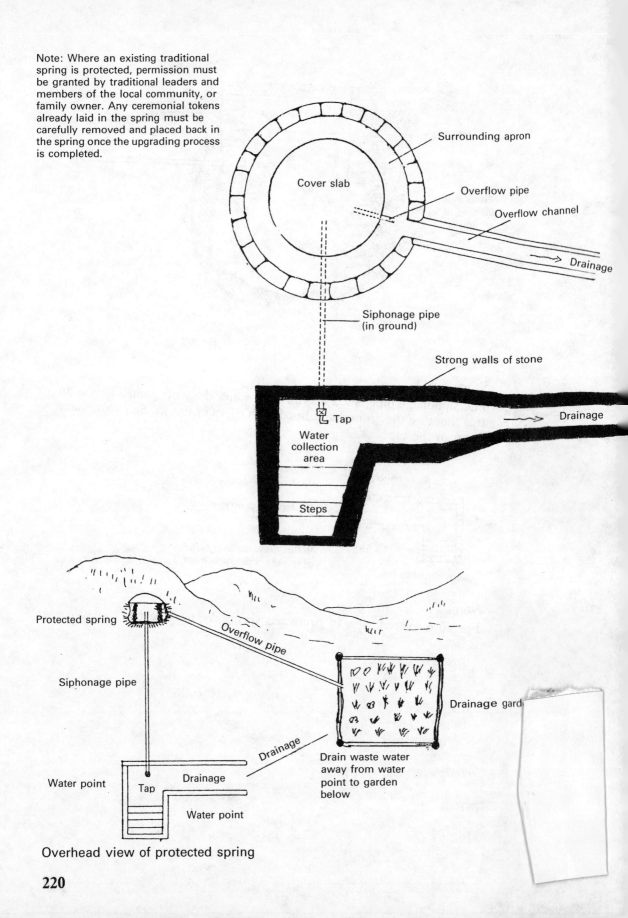

Overhead view of protected spring

The siphon well

The siphon well is a little known but very valuable system which is useful in areas where the well tapping groundwater occurs at a slightly higher point than the living area of the community it serves. The water level in the siphon well can in fact be some metres below the ground level at the point of the well itself, but this water level is above the point where the water will finally be taken.

The diagram (on page 222) shows the position and layout of a typical siphon well. The well is dug and protected using the most appropriate method, as described in earlier chapters. A channel is cut as deep as possible from the lower side of the well to the site where the water point, usually a tap, will be placed. A pipe is now laid in the channel from the apron on which the supply tap is to be mounted, through the side wall of the siphon well and down into the well water. Preferably, the pipe which passes through the wall of the well should pass a point below the resting water level in the well.

The pipe which passes through the wall of the well should be made of strong material, either steel or heavy-duty polypipe. It is very important that the siphon pipe is completely airtight. There is one particular problem with the siphon well that must be overcome — that is the possibility of developing an airlock — in which case the water will not run.

Air locks can happen if air is allowed to run up from the 'tap' through the rising pipe, so that air accumulates in the highest part of the siphon pipe, which will be where it enters the well. Air locks can also develop if the well is allowed to run until the water level reaches the lower end of the suction pipe in the well. Air will then be drawn up and the siphonic action will stop.

Both of these problems can be avoided by following a certain design plan. First it is essential that the complete pipe is airtight. Secondly, the joint between the tap unit and the lower end of the delivery pipe should be airtight. Air will have great difficulty in travelling downwards through the vertical member of the tap pipe. To avoid the well water being used until the siphonic action stops, an inverted ball valve can be fitted on the end of the suction pipe. In this case the ball valve itself will open when there is an ample depth of water. As the water level drops closer to the end of the suction pipe, the ball float of the ball valve will begin to drop and close off the valve, thus limiting the amount of water travelling through the system. The valve will close off completely before the water level reaches the water inlet in the valve.

Starting the siphon

Ideally the resting water level in the well should be above the level at which the water delivery pipe enters the side wall of the well. If this is the case the piping can be installed at the same time as water from the well is being bailed out and thus is at a low level. If the system is left overnight the water level will rise above the water outlet and the siphonic action will begin as the pipe is now filled with water.

In the case where the water level is too low for this technique to be used, it will be necessary to pump water up through the tap under pressure to exclude all air from the siphon pipe. Bubbles of air will be seen rising from the bottom of the suction pipe in the well. When all these have been excluded it will be possible to try the siphon. Once it has been started and the siphon pipe is free of air, the water should run freely. The best siphon pipes are made so that they are without joints. A good example of this would be to run a heavy-duty polypipe down into the well and fix it to the side wall, running the other end to the tap outlet. Heavy-duty polypipe is perhaps the best material to use.

Of course if the well is one which yields a great deal of water, it would be possible to allow the water to run freely continuously — rather like a protected spring. Most protected springs are fitted with taps to build up the static level of water in the 'spring box' — but some springs are so generous in their water supply that they can be left running continuously. This can also happen with siphon wells. However if there is any doubt about the potential of the water supply, taps should always be fitted as this conserves the water in the aquifer feeding the well.

A siphon well can provide long and very reliable service — and may run without maintenance for decades. Several have been built in the Midlands Province — and one built for the Ministry of Health camp at the Henderson Research Station has delivered high quality water for over a decade without needing attention.

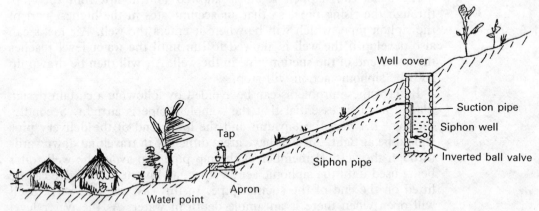

The siphon well.

The gravity well

The technique described in this section is related to both springs and siphon wells. This system is similar to the siphon well, water runs from the well to the water point directly by gravity. It is therefore referred to as a 'gravity well': there is no siphonic action.

The first stage of the technique involves digging down deep into the well so as to form a hole which penetrates the water table. In this case the water table was well above the area where people were living, and thus by suitably protecting the well it was possible to gravitate the water to the living area. The main operation involves digging deep into the water-bearing layer, which in the example described here was made of very sandy soil. In addition to the main hole being dug a deep trench was dug from the site of the well to a lower point. The delivery pipe was laid in this trench. In one example the distance from the surface to the water was 1.5 metres; the surface to the bottom of the well was 2.4 metres, the delivery pipe being 2.15 metres from the surface, allowing approximately 0.25 metres of water in the base of the well below the pipe. The trench was dug slightly below the pipe level in such a way that the pipe always sloped downwards away from the well.

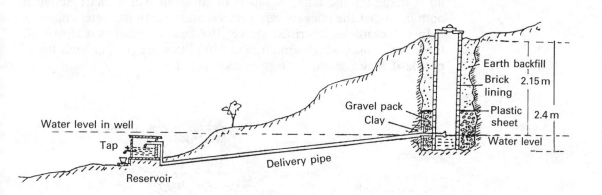

Cross-section of a gravity well.

Obviously it is desirable to dig the well and the trench leading from it as deeply as possible and this requires a great deal of effort at the time but is certainly worthwhile.

Once the hole is dug as deep as possible into the water-bearing layer and the trench dug to almost the same depth at the point where it enters the well (0.25 metres above the base of the well), it is possible to start laying a circle of bricks without cement mortar about 1.2 metres in diameter (4 ft) and one metre deep. These are laid in the centre of

223

the well excavation which should be at least 1.5 metres in diameter. This allows for the introduction of a gravel layer around the brick work, known as a gravel pack. The gravel, which can be anything from sharp river sand to 25 mm chips, is poured inside the space between the earth wall of the well and the brickwork. The delivery pipe is placed through the brick wall about 0.25 metres (9 inches) above the bottom of the well. This is best made of 25 mm steel pipe or heavy duty polypipe. Above the water level, the bricks can be mortared together. The gravel packing is continued until it lies above the water layer, when a layer of plastic sheeting can be laid over it. The mortared brickwork is continued until it lies about 300 mm above ground level. The annular space between the wall of the well and the brickwork is backfilled with the cuttings from the well and soil, which is packed well down. A concrete cover is made for the well and a concrete apron laid around this, at least 2 metres in diameter.

The delivery pipe is extended to a lower level, so that it slopes continuously downwards. At a point well below the water level in the well, a brick tank is made and the delivery pipe is fitted so that it allows water to run into the tank, as shown in the diagram. It is important that the water runs freely into the tank under the force of gravity. A low pressure ball valve is fitted to the inlet pipe, to close off the inlet pipe once the tank is full. This is the same type as used in a flush toilet cistern. All exposed piping should be made of steel, with piping fitted under the ground made of heavy duty polypipe or thick-walled PVC.

The tank can be made about one metre in diameter and 1.5 metres high, the inlet pipe coming in close to the top of the tank. A concrete lid is made for the tank. A tap is fitted about half a metre above the bottom so that the buckets can be fitted underneath for water collection.

In the example described above, the flow of water was up to 3.75 litres per minute which amounts to 5400 litres per day, enough for 270 people if they consume 20 litres each per day.

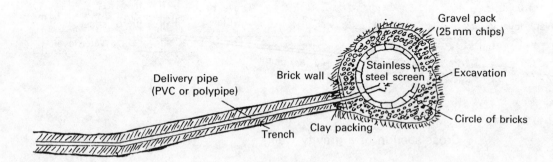

Plan of a gravity well.

Rainwater harvesting

The collection and storage of rain from run-off areas such as roofs, rocks and other surfaces has been practised by Man since ancient times. It is still used in many parts of the world and is particularly suitable for areas where pumped or reticulated supplies of water are not available. It is a technique that has been tried in several parts of Zimbabwe but utilisation of this excellent technique has not been fully exploited. By careful design it is possible for a family to live for a year in areas with as little rainfall as 100 mm per year.

Many observations made in Zimbabwe show that between 80% and 85% of all measurable rain can be collected and stored from suitable catchment areas. This includes light drizzle and dew and condensation which can occur in many parts of the country during the drier months.

As an example, if the rainfall is 635 mm per year (that in Harare is 900 mm/yr) the run-off from a suitable catchment area will amount to approximately 500 mm. One millimetre (mm) of rain falling on one square metre of roof will yield approximately one litre of water. 500 mm of rain will therefore yield 500 litres of water from one square metre — measured horizontally.

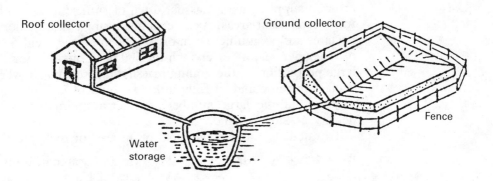

Water harvesters.

Contamination

All harvester surfaces, being exposed throughout the year, are subject to contamination by dust, insects and birds and those at ground level are also liable to be contaminated by animals and humans. Ground harvesters should be properly fenced and kept clean — the first flush of the new rains should be run to waste. Storage tanks can be built either below or above ground, those below ground should be fitted with a handpump. The tanks should be fully enclosed to prevent evaporation. All apertures should be screened to prevent the access of mosquitoes, insects and rodents, lizards and other life. Where this is done the only source of contamination will be the roof or collection area. It may be necessary to pass the water through a sand filter before it is consumed for drinking — especially if there is some doubt as to the purity of the water. Several tanks taking water from roofs and built in the Epworth area have been tested by Blair Research Laboratory staff for faecal *E. coli*, and results are consistently low for this source rarely exceeding 10 colonies per 100 ml sample. This type of source can therefore be considered safe for drinking provided care is taken to cover the tank and screen the inlet.

Consumption rates for water

If we assume that water is to be consumed solely for domestic use, i.e. for drinking, cooking and washing, each member of the family may consume between 15–20 litres of water per day. If we assume the family size is six and each person, including children, consumes an average of 15 litres per day, the total consumption for the family is approximately 100 litres per day. If we assume that the longest period without rain will be 6 months (180 days) the volume of water required to last through the dry season will be $180 \times 100 = 18{,}000$ litres. This will be the minimum size for any storage tank built for this purpose.

Types of harvester

The granite areas of Zimbabwe which cover over 50% of the country are well supplied with massifs, often of sufficient size to be utilised as water catchment areas. As an example a single granite dome covering 1 hectare, and assuming a run-off of 500 mm may yield 5,000,000 litres of water per year — or enough for 500 head of cattle for six months. A large proportion of the granite massifs in Zimbabwe cover an area of 1 hectare or more and, if fully utilised, could play a very significant role in providing the large number of watering points necessary in a high density grazing scheme.

To calculate the average yield of an area of rock use the formula:

$W = 4/5 \times A \times R$ where W = yield of water in litres
A = area of harvester in sq. metres
R = average rainfall in mm. and
4/5 is the run-off factor.

Roofs

These may be particularly useful for domestic use in the private homestead or at schools where modern roofing material is used. Thatch does not really lend itself to rainwater catchment although it has been used in some areas. For tile, asbestos or tin roofs an average farm house may have a roof area (including outbuildings) of 200 m^2. If the run-off is 500 mm/year, the yield will be 100,000 litres per year. If adequate precautions are taken against contamination, the water should be soft, potable and suitable for washing. The same formula can be used as above, the roof area being calculated for the horizontal area.

Water is led from the roof to the storage tanks through a series of gutters and pipes. Conventional gutters are normally used, but these can be made more economically with V-shaped lengths of tin sheet hung under the roof edge from wire, or lengths of PVC pipe cut along the length and clamped to the edge of the roof.

The most useful place for the roof rainwater collection technique is at the rural school. Usually these are built with extensive roof coverage, and large quantities of rainwater can be collected through the season. Careful management of the water, essential to the success of the technique, can also be taught at schools.

Artificial harvesters

If water is required where there is no roof or rocky outcrop suitable for collection, the construction of an impervious surface can be undertaken on the ground itself. Reinforced concrete can be used as a surface — chicken wire reinforcing should be used to prevent cracking. An alternative technique is to lay a large piece of plastic sheeting in a hollowed out and levelled area of ground. A layer of sand is laid over the bottom of the excavated area and raked flat. The sheet of plastic is laid out over the layer of sand and the edges raised up against the side walls of the excavation. A drainage system is now laid in the catchment area on top of the plastic sheet in the form of a slotted PVC pipe which drains away into the reservoir used for storage. Finally a layer of gravel or very coarse washed river sand is laid on the bed. The edges of the area should be raised with a rim of concrete work. No part of the plastic sheet should be exposed to the sun or it will perish within a year. The area should be fenced off to prevent access to animals. To calculate the area required for the water requirement use the formula:

$A = 5/4 \times W/R$

If a family wishes to collect 20,000 litres of water per year the size of the harvester should be $50 \, m^2$.

One method for harvesting water in a homestead from the ground.

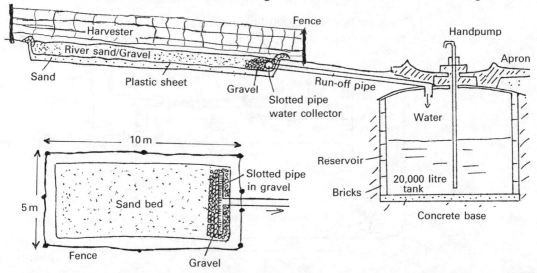

The system illustrated uses many ingredients, such as bricks, sand and stone found in the communal lands. Plastic sheet, PVC pipe and cement must be purchased and imported, however. A suitable hand-pump can be fitted over the reservoir if it is built underground.

Reservoirs

The most practical method of using rainwater is to collect it from a roof or series of roofs and run the collected water through guttering and piping to a tank or series of tanks. Where roofs already exist, the greatest cost is the construction of suitable tanks. Where the potential for collecting large volumes of rainwater exists, tanks should be built large enough to cope with the volume. If they are built well in concrete or cement mortared brickwork, they can be considered a worthwhile investment. Ideally the tanks should be covered, which prevents the growth of algae, reduces evaporation and helps to prevent contamination.

Where roof or rock harvesters are being used, reservoirs can be built above ground level. However in some artificial harvesters and some rock harvesters the elevation is very low and the reservoir must be built below ground level. Two basic methods are used for building tanks. One system uses burnt bricks for the walls, the other reinforced ferro-cement.

Building tanks

1. Brick tanks

A 20,000 litre tank is a very convenient size, both as a unit for a homestead and also as one of a number of units that might be built at a school. A 20,000 litre tank should be approximately 4 metres in diameter and 1.6 metres high, and can be built with 110 mm (4½ inch) thick (single course) brickwork. Brick reinforcing wire should be included in the mortar between the brick layers.

A strong base slab is first laid 100 mm thick, the mixture being 4 parts stone, 2 parts river sand and 1 part cement. This is made 600 mm wider than the internal diameter of the tank (4.6 metres). The base slab

Cross-section of 20,000 litre brick tank.

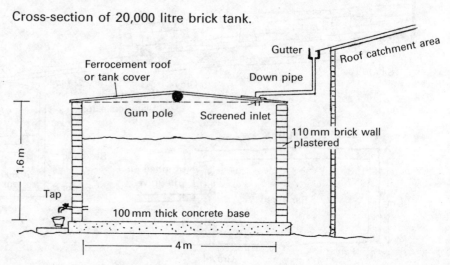

is reinforced with 6 mm steel reinforcing bars laid at 150 mm intervals in a grid pattern. This is allowed to cure over the course of a week, being kept wet to develop strength.

The brickwork is now laid in a circle, building up one course at a time from the base of the tank as shown in the diagram. Reinforcing wire which can be 'brickforce' or fencing wire, is laid between each course to give strength. A steel pipe fitted with a tap is introduced through the brickwork, as shown in the diagram at a point which is as low as possible, but will allow a 20 litre bucket supported on a concrete base to stand beneath it. The brickwork is built up to the required height, which is about 1.6 metres. The internal surfaces are plastered with strong cement mortar and steel floated. The addition of a waterproofing material called 'impermo' helps to seal the plaster.

The tank cover is now made. This can be made with tin or asbestos roof sheeting supported by a wooden beam placed across the middle of the tank. It can also be made with ferrocement. In this method a central wooden beam, which can be a gum pole, is laid across the tank and a piece of 25 mm chicken wire cut to suit the tank. The wire is removed and a piece of hessian is cut to the same size and attached underneath the chicken wire. The attachment can be made with thin wire or string in many places so that the hessian is well supported all over. The wire/hessian sheet is then laid back over the tank and supported by the pole and anchored to the side walls of the tank. A cement slurry is now made with 3 parts of sand (mixture of pit sand and river sand) and 1 part cement. The slurry is applied to the surface of the wire/hessian with a brush, so that it soaks through both layers. The following morning another layer is added and this is allowed to cure. As with all cement work, it should be kept wet at all times, after it has set. The

A 20,000 litre water tank can be made from bricks using a single thickness of brickwork if wire reinforcing is used between the brick courses.

slurry layer should be capped by an additional layer of plastered cement mortar to bring up the thickness to 25 mm. The wire should be well embedded in the cement mix.

Before the final plastered layers of the roof are added, a water inlet hole should be cut into the ferrocement at the most suitable point. The inlet hole can be built up with mortar. It is best that the inlet hole is fitted with a stainless steel screen.

2. Ferrocement tanks

A variety of methods are available for the construction of ferrocement tanks. These are built with steel reinforcing wire and cement mortar.

Method A

This is a technique which can be used to make tanks up to about 5000 litres using concrete, weldmesh, hessian and wire.

In practice it is convenient to buy one place of weldmesh (6 m × 2.4 m) with a wire thickness of 3.55 mm with 100 mm squares. From this it is possible to make a tank 1.5 metres high and slightly less than 2 metres in diameter, the base being slightly more than 2 metres in diameter. The standard piece of weldmesh is cut so that one piece is 6 m × 1.5 m. This piece is used for the walls of the tank. The remainder is cut up for reinforcing the base of the tank. Hessian is sold in 1.38 metre (54 inch) widths, which is suitable for the walls.

The base is made first, the site being levelled and a ring of bricks laid on the ground as a mould, 2 metres in diameter. The weldmesh is now cut and the long 6 m × 1.5 m piece bent into a tube — the ends being wired together. This tube will be 1.5 metres high and 1.85 metres in diameter and is used to support and reinforce the walls. The remaining steel is cut into 3 lengths of 2 metres × 0.9 metres and laid within the brick mould and shaped or cut so that the wires fit within the bricks.

Next a mixture of concrete is made for the base. This is made of 4 parts stone, 2 parts river sand and 1 part cement, and requires about 2 bags of cement. Half the mixture is a put inside the mould and the wire reinforcing is placed in position. The remaining concrete is now added and trowelled flat.

After the concrete has been poured and levelled a tubular-shaped piece of weldmesh is inserted in the base, so that the lower wires are embedded well in the concrete. The weldmesh frame is built up in a circular wall to the desired height (1.5 metres). Lengths of heavier steel rod are then wired around the wall of weldmesh. However before the heavier steel (6 mm) wires are mounted, a suitable length of hessian is stretched around the weldmesh wall and tied tight with binding wire. The reinforcing wires are then attached so that the hessian is placed between the wire layers. A slurry of sand and cement (3:1) is then mixed with water and flicked on to the hessian with a large brush. The whole area of the wall is covered on both sides and built up to about 10 mm thickness. This slurry is left to cure overnight.

The application of plaster now begins. A plaster coat consisting of sand and cement is now mortared on to the skin of slurry, this mixture being 5 parts sand and 1 part cement. (The sand is pit and river sand mixed in equal proportions.) It is possible to wind extra lengths of wire around the plaster coats to provide extra strength. The plaster layers are built up to a thickness of between 25 mm and 40 mm. The inner surface is steel floated so that it becomes more watertight.

It is possible to make a roof by obtaining some more weldmesh and cutting this to a suitable shape. A layer of hessian is now stretched on to the weldmesh and laid so that the mesh is uppermost and the hessian below. The two layers must be joined with many pieces of thin wire. Before the roof is fitted a final layer of cement mortar is laid on the inside surface of the tank and steel floated to make it water tight.

If a tap is required — and this will normally be the case if it serves as a roof catchment tank — a hole is knocked into the tank and a tap together with steel socket and barrel nipple is securely concreted into place, as shown in the diagram.

A tube of weldmesh is mounted in a base of concrete, it is covered tightly with hessian and a cement slurry is flicked on to it with a brush, a further layer is applied in the same way. Once the slurry has set, layers of cement mortar are added with a wooden float, the final layers are steel floated.

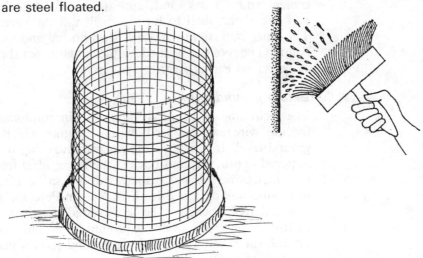

Once the plumbing and interior surface of the tank have been finished, the roof can be placed in position and secured with wire. More slurry is now added to the roof as it was added to the walls. Once the slurry is hard, a plaster layer is added for strength. Once finished the entire tank can be filled with water. A hole is made in the roof to allow water to pass in from the rainwater catchment area. A stainless steel screen should be added to this opening.

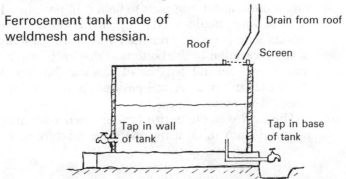

Ferrocement tank made of weldmesh and hessian.

Method B

Above ground tank

The second technique is similar to the first, the difference being that a layer of chicken wire is placed around the weldmesh walls prior to plastering. The technique involves stretching suitable lengths of 25 mm chicken wire around the weldmesh (100 mm spacing) walls and then strapping a series of reed matts (or hessian) tightly around the outside of the wall. Once these have been placed in position and tied with cord or wire, the application of mortar can take place from within the tank. The mortar can be made with a mixture of 1 part cement and 3 parts sand (half pit and half river sand). This is plastered on to the wall so that it penetrates the chicken wire and weldmesh and forms a coat against the reed matt or hessian. A second coat is added to the first, two hours after the first application.

This initial layer is allowed to cure for one whole day, then the reed matt or hessian is removed, forming a shell on to which further layers of cement mortar can be laid. In cooler weather two days should be left for curing. These outer layers can be made with a mixture of 1 part cement and 4 parts sand, and are added on to both inner and outer surfaces of the shell to form a final wall thickness of 40 mm−50 mm. The inner watertight surface is steel floated and the plumbing added as before. The roof can be made with standard roof sheeting or ferrocement as described earlier in this chapter.

Below ground tank

The technique of using chicken wire in combination with a stronger fencing wire can be used to make tanks which are partly beneath ground level. In this case the excavation is dug into the ground to the required depth, a round bowl shape being ideal for strength. Tanks up to 6 metres across and 3 metres deep can be made in this way. The technique was designed by Erik Nessen-Peterson in Kenya.

A collar of bricks can be laid around the rim of the excavation and built up about 300 mm above ground level. This collar can be extended up to form a wall a metre above ground level if necessary, especially if the area of water collection lies above the tank position, when greater tank volume can be obtained.

The bowl-shaped excavation is then painted with a mixture of equal quantities of cement and water, the application being made with a brush. Next day the entire surface, including the upper brickwork is lined with a 30 mm thick layer of stiff cement mortar (1 part cement, 3 parts sand). This is left with a rough surface.

The next day the entire excavation should be lined with the chicken wire reinforcing, using nails to hold it in position. The various pieces of chicken wire should overlap by about 150 mm. Next, a long length of fencing wire is nailed over the chicken wire spirally down from the top of the excavation to the bottom so that each loop is 150 mm beneath the one above. Several lengths of fencing (barbed) wire should be laid down the tank in a vertical position. Some cutting and rewiring will be required to obtain a neat fit.

The next day the entire tank is lined with another layer of cement mortar 30 mm thick using the same mixture as before. The mortar is thrown on and trowelled smooth with a steel float. The mortar should extend up any brick wall that is built at the surface. A final coat of liquid cement (1 part cement + 1 part water) is then trowelled on to the mortar. When it is finished a layer of plastic sheeting should be laid over the mortar to keep it damp. The concrete work should be kept wet at all times to assist curing.

The roof can be laid over the surface. The type of roof may depend on the size of the tank prepared. A conical shaped roof is recommended for larger tanks by UNICEF, P.O. Box 44145, Nairobi, Kenya, from whom complete plans can be obtained for ferrocement tanks. With smaller tanks up to 4 metres across, treated gum poles can be laid across the top of the tank and covered with tin sheeting or ferrocement/hessian sheeting as described earlier.

Tanks of this type can be built in series with an outlet from the first leading to the inlet of the second a little further downhill. A third tank can also be built if necessary.

These tanks are useful for storing water from rock dome outcrops, which can serve as excellent rainwater collection areas. A low brick wall is built between 150 mm and 300 mm high around and up the granite dome in such a way that it collects and diverts the flow of water from the dome into a channel which leads to the tank. The wall is built up in bricks and mortar from the rock to form what might best be described as masonry gutters attached to the rock face.

Water collected in reservoirs built below ground level must either be pumped out or siphoned out through piping to lower lying areas.

Ferrocement tank plastered below ground level.

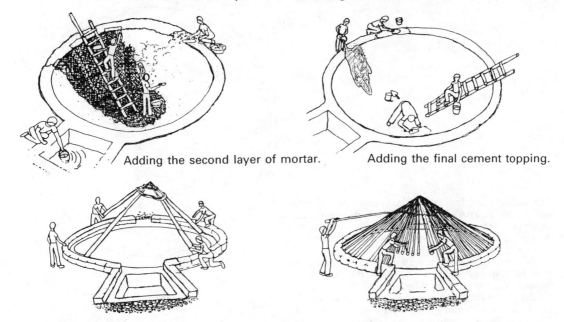

Adding the second layer of mortar.   Adding the final cement topping.

One method of adding a roof with wooden poles.

This gravitational and storage system was designed by Erick Nessen-Petersen (Danida-Kenya).

233

Method C

This technique uses a corrugated iron mould made for the job. In this case each mould consists of 4 segments which are bolted together to form a tank shape. Each segment is made so that it is fitted on to rigid iron end pieces.

First a concrete base is made as in previous tanks. Second the corrugated iron mould is erected. This mould can be made in several sizes. The one shown in the diagram is 1.8 metres in diameter and 1.6 metres high.

Notice that each segment is separated from the next with a series of spacers made to leave a gap of about 50 mm. The spacers are made with a suitable length of 20 mm steel pipe — the segments being secured together with steel nuts and bolts. Before the chicken wire is stretched around the mould a strip of 'damp course' used in brickwork technique, is fitted against the spaces left in the mould. The chicken wire is then stretched around the mould and this is reinforced by stretching bands of 8 gauge wire around the mesh tightly. Cement mortar is now added as before, left to set and another layer added. After two days the mould is removed by undoing the bolts and separating the segments. Additional layers of mortar are added both inside and outside to the required thickness. The plumbing is made as before.

The roof can be made with roofing sheets or with ferrocement as described for earlier tank designs.

Mould method of making a ferrocement tank.

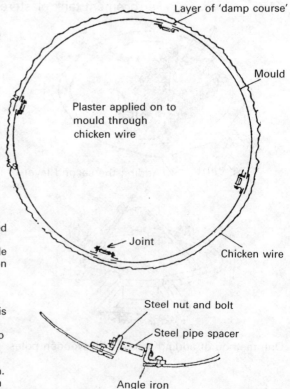

This system uses four curved corrugated iron sheets which form the sections of a mould which when fastened together make a complete circle. The upright edges of the corrugated iron sheets are riveted to lengths of angle iron for strength and as a means of attaching one section to another, the sections are attached to each other by steel bolts spaced with a pipe. The 50 mm gaps left between each section are filled with a 'damp course' during the construction stage. A layer of chicken wire is tightly wrapped around the mould and secured with 3 mm wire loops. Cement mortar is applied in layers to the mould through the wire, once the concrete has cured, the mould can be removed from inside by undoing the nuts and bolts and removing each section. Further layers of mortar can be added to provide extra strength.

Water point design

The provision of water from protected springs and rainwater catchment tanks is usually made through a tap, and in order to preserve a hygienic environment it is important that the tap and its surroundings are well mounted and surrounded by an adequate apron and waste water run-off system that will adequately dispose of spilled water.

The water point can be considered in four equally important parts:

1. The water delivery system
2. The apron
3. The water run-off channel
4. The waste water disposal system

The water delivery system

Normally, a conventional heavy-duty screw type tap can be used on the end of the delivery pipe, which should be mounted in concrete. Where the amount of supplied water is more than adequate for the needs of the community, a slight loss of water by wastage may not be a serious matter. In this case the conventional tap is adequate for the job, and is the most commonly used. However if the source of the water is limited, or if water is pumped at some expense, certain measures may be required to restrict the wastage of water. A design of a modified tap system is described below and has been on trial by Blair Research Laboratory for about ten years. It is illustrated overleaf.

This uses a press tap mechanism, mounted at ground level and foot operated. The water delivery spout is short and only high enough to receive a 30 litre vessel. The body of the tap unit and the water delivery pipes are mounted within a central block of concrete. The stud of the tap itself is designed so that the working parts can be removed with a spanner to facilitate maintenance, such as replacement of washers etc. Some modern self-closing taps are complex and employ a delayed closing mechanism. A spring-loaded tap may be adequate provided it is a heavy duty type and correctly mounted. Taps of this type do employ water seal washers within then, and it is essential that the mechanism can be opened up to facilitate a complete repair service.

Taps of this type are made of brass with neoprene water seals. The button of the tap is best fitted at an angle of 45 degrees as shown in the diagram. The foot operation is then easy, but it is more difficult for users to keep the tap running by means other than the foot.

The water delivery pipe can be made of 20 mm or 25 mm galvanised piping with a delivery which aims water downwards into the waiting bucket. The working parts of the tap/delivery system are cast in high strength reinforced concrete and cured under water. The thread of the tap (20 mm) is left exposed for fitting to the mains supply. A female socket is also cast into the concrete so that the water delivery pipe can be fitted to the appropriate height when the installation is being fitted.

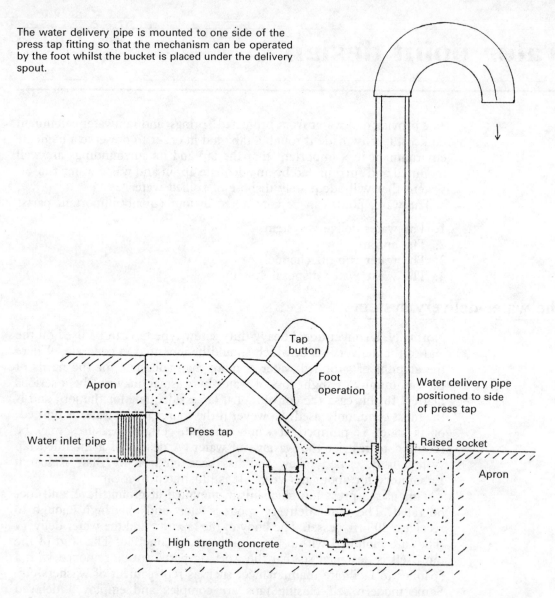

Water delivery system.

The apron

It is assumed that an appropriate site for the water point will have been chosen and also that a piped water supply (from a spring, water tank or other source) will have been laid on to this site. Clearly the site should be convenient for people's use and also in an area which is not prone to swamping. It should be slightly elevated so that run-off water can drain away from the water point.

The water delivery unit (or standpost) is plumbed into the piped water supply, and then the apron is built around this central unit. The apron should be made of reinforced concrete and should be at least 2 metres and preferably 3 metres in diameter with raised edges. All waste water should drain towards the water run-off channel.

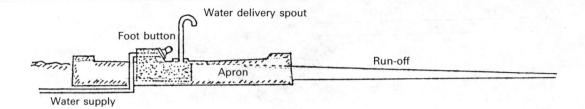

Commercially-made moulds are available for the construction of aprons and water run-off channels (V & W Engineering, Harare.).

The water run-off channel

This is an open channel made of well-mortared brickwork or reinforced concrete. The base of the channel should be at least 300 mm wide and have a total length of at least 6 metres and preferably 10 metres. The channel should be built with a slope so that all the water entering its upper end will drain away completely to the lower end.

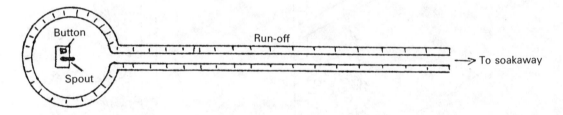

The waste water disposal system

This is perhaps one of the most important parts of the complete system, and apart from the tap itself, is the most likely to give trouble. There are various methods of disposing of waste water. These can be listed as:

1. Soakaway method
2. Evapotransporation method
3. Sump method

1. Soakaway method

This is the most common method of disposing of waste water, but is not always the most successful because of the high content of silts and solids which find their way down the run-off channel. Soil is carried on to the apron on the feet and is also used for washing buckets. This material is washed down the channel and builds up in any depression, including the soakaway. Often the ground level around an apron falls because soil is taken away from it, and builds up in the soakaway.

Several types of soakaways are available for use. The diagrams overleaf show some of these.

The volume of these soakaways should not be less than the expected daily discharge. Design 2 is far more efficient than Design 1. In this

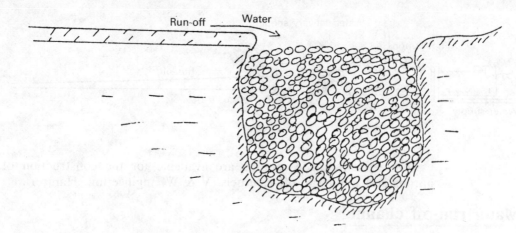

Design 1. Simple soakaway filled with stones: this type will eventually block.

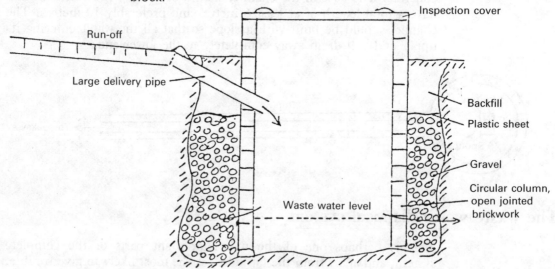

Design 2. Open jointed brick tube and gravel pack type.
The inspection cover can be removed to take out solid accumulation.

case the pit is best constructed with a tubular-shaped cylinder of open jointed brickwork or stones lying centrally within it. A layer of broken rocks or stones at least 300 mm deep should be inserted between the brick wall and the earth wall. The cover slab allows for periodic cleaning of the accumulated solids.

An alternative to the first design is a soakaway trench. In this case a 100 mm diameter pipe is led from the run-off through the ground within a trench which may be 15–20 metres long. The trench is dug 1 metre deep and 0.5 metre wide. The pipe is laid within this and surrounded by crushed rock or gravel. The pipe is open jointed to allow effluents and waste water to flow out into the gravel layers in the trench. The gravel layers are covered with thick plastic sheet and backfilled with soil. It is best to have a sump box between the open water run-off and any underground piping to trap any solid matter.

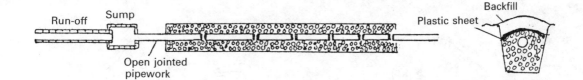

2. Evapotransportation method

In this technique waste water running off from the run-off is allowed to travel down a drainage furrow lined with plants which take out the water from the saturated soil.

Plantations of banana, sugar cane, gum trees and even vegetables in a garden serve adequately for this purpose. Even areas of grass can serve this purpose well if properly maintained. In fact a fair degree of maintenance will be required whatever the plant used. This includes the removal of sediments and other wastes washed down and accumulating in the seepage area. The illustrations show a typical transportation area and also a system using banana plants. These must be cared for when young, but will cope well when matured.

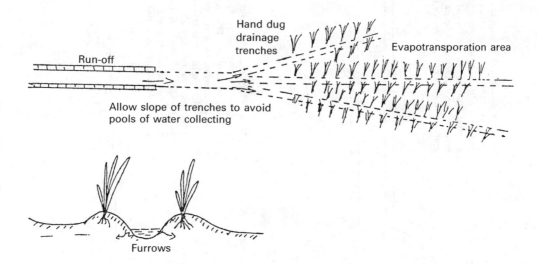

3. Sump method

In this case the waste water is led into a vegetable garden owned by a school or family. Water is led from the run-off directly into a sump which lies within the garden area. Water is withdrawn regularly from the sump with a bailer or bucket to water the vegetables which surround it.

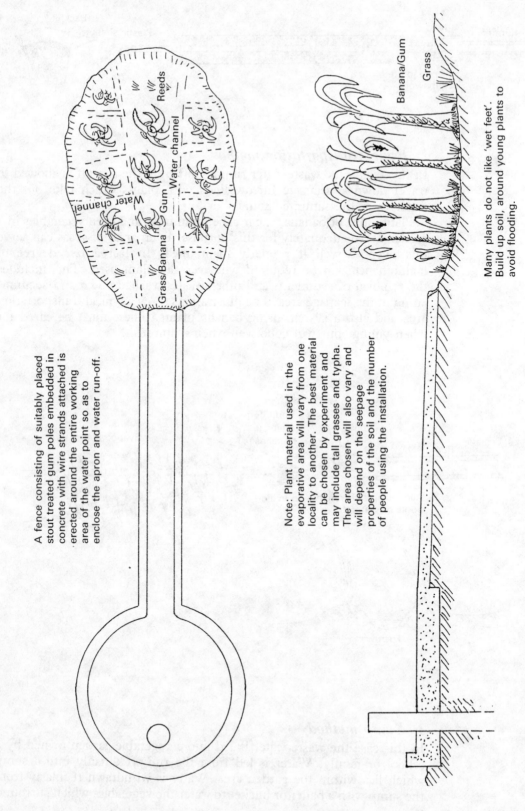

Use of banana plantation as seepage area for water point.

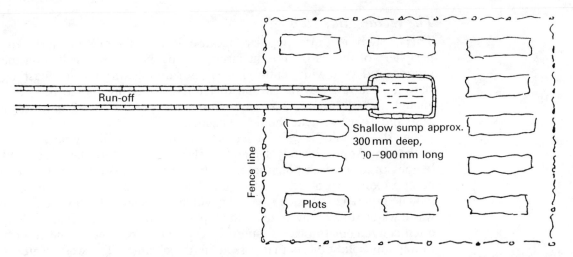

Base of sump can be left unlined to allow for seepage into ground if water is not bailed out into garden.

Simple washing slab design

The handwashing of clothing is an important part of daily life in most households. Traditionally, washerwomen very frequently wash their clothes in a river if this is nearby, where rocky outcrops often form suitable washing surfaces close to where water is available in abundance. The fact that simple flat surfaces available in nature are found to be both practical and acceptable means that artificially-made slabs made of concrete do not need to be sophisticated in order to be workable.

Many descriptions made of washing slabs show complex raised platforms with sinks and ridged surfaces — designed so that they can be used standing up. These are being used closer and closer to the site of remote water pumps in the communal lands.

However, much simpler techniques can be used with success. A simple, suitably-shaped concrete surface provided near a tap or water point and served with a water run-off channel and water seepage area is all that is required in the first instance.

The round slab

The round slab should be made about 3 metres in diameter, and the rim should be raised and the surface of the slab sloped so that water will freely run off into the water channel and seepage area. A central water point such as a tap is located as shown in the diagram. This permits water to be taken off centrally, whilst the washerwomen kneel at the rim of the slab.

Circular washing slab.

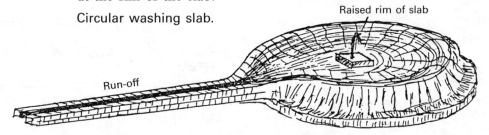

The rectangular slab

This system is more popular because it offers more room on the periphery of the slab. Like the circular slab, the concrete is laid on the ground and cast with reinforcing wires running through it. Washing slabs do wear away with use and the concrete mixture should be strong enough to last several years. A mixture of 4 parts stone, 2 parts river sand and 1 part cement should be adequate. This should be well cured.

The slab should be constructed around the water delivery point, such as the tap. Approximately 1 metre of slab per peripheral length should be measured for each potential user. A slab for 12 people will therefore be approximately 6 metres long and 2 metres wide. This will accommodate 6 persons on each side. The whole slab should be sloped so that waste water runs off away from the tap and into the water channel. It is often convenient to place a 'sump box' at the upper end of the run-off to trap any solid objects that pass along the channel. The waste water is led to a soakaway or seepage areas as described earlier.

The best way of building the washslabs is to build up the walls of the slab with bricks and cement mortar, and then backfill the area within with reinforced concrete and stones and bricks. The surface is smoothed off as shown in the diagram. A washing line can be placed near the slab so that clothes can be hung up to dry in the sun. This need be little more than a series of posts anchored in the ground with a strong galvanised steel wire suspended between them.

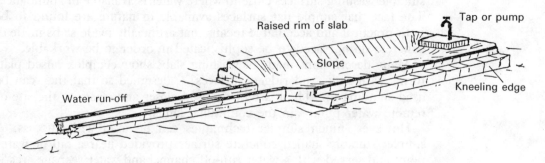

Rectangular washing slab.

Washing slabs of this type are known to attract people away from river banks, and thus reduce the potential for bilharzia transmission in riverine areas.

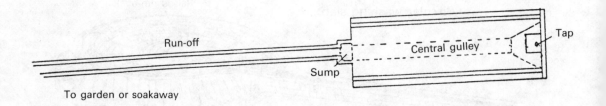

Washing slabs used near Bush Pump water points

This is a more complex design than the simpler washing slabs described earlier. It is a popular design however, and many are being erected near Bush Pumps, which may be sited some distance from where people live.

Detailed descriptions of this design are available from the DDF.

Hygiene aspect of water supply

Hygiene in the homestead

1. Storage of water

If the full benefit is to be derived from any improved water source in the rural areas it is essential that the water is collected in a clean bucket and stored correctly in a vessel with a covered lid.

In surveys carried out on the quality of water tested from a source and from the containers used to collect the water, it has become clear that the containers themselves may often be the source of some contamination. Whilst it is impossible to expect these containers to be sterilised, health workers should advise families that the water container should be used for no other purpose, and that it should be kept clean at all times and when water is stored in the vessel, it should be covered and kept out of the way of very young children, who may contaminate the water.

2. Hand-washing

Improved health can never directly result from the introduction of improved water supplies alone. It must be accompanied by a hygienic use of the water itself to have any effect on the health of the individual.

It is also true that no individual can possibly benefit from the hygienic practices associated with water unless water is available, close by and in reasonable quantities.

Thus the road to health as far as water is concerned must take place in two stages. First the water must become available, close by and in reasonable quantities; second the water that is available must be accompanied by hygienic practices by the individual to have an effect on the health of the individual. One of the most important of these is body washing and hand washing in particular.

The hands are possibly the most mobile parts of the body. They touch the ground, doors, taps, towels, other peoples bodies and they are used to clean the body they belong to. Of all the parts of the body they are very susceptible to contamination. The hands may frequently pass from one place to another and often into the mouth, especially in children. It seems logical to suppose therefore that the hands may be responsible for carrying pathogenic bacteria — from contaminated places directly to the mouth — especially in young children — who seem to suffer most from enteric disease.

Clearly the washing of the hands is a most important exercise after a person has made a visit to the toilet. But how often is there water close at hand?

Ideally every latrine constructed in a family setting or anywhere else should also be fitted with some form of simple hand-washing facility.

The mukombe

One of the simplest and most elegant hand-washing devices was designed by Dr Jim Watt of the Salvation Army in Chiweshe. This simple device is cheap to make, effective and economical in its use of water and has been called the 'mukombe'. In its simplest form it is the 'mukombe' fruit that is taken straight from the land from a trailing plant. Often it is dried out and used as a water bottle, cup or gourd.

The mukombe.

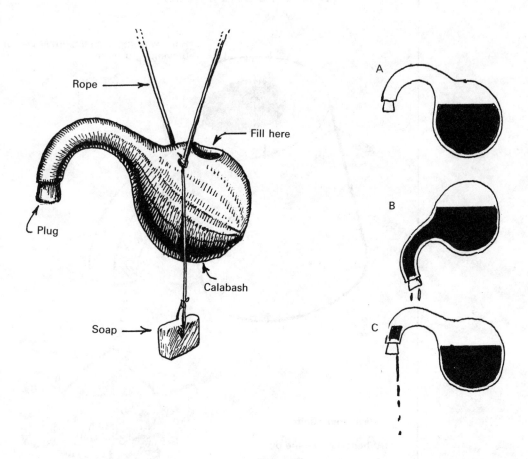

The mukombe is the fruit of an indigenous plant and can have many uses in the rural setting of Zimbabwe. It is often used as a cup or spoon. It is very common in many areas of Zimbabwe and can be formed into a hand-washing implement very easily. The idea is very simple and elegant and was first demonstrated by Dr Jim Watt of the Salvation Army in Chiweshe. Many vessels can also be used in the same way. What is important is that people have a simple means to wash their hands easily.
Illustration by Jim Watt.

A leaflet used in campaigns to promote handwashing in rural communities. The mukombe is a suitable vehicle on which to promote campaigns of this type. It removes bacteria from the hands very efficiently with small amounts of water and is in fact more effective than the method used with bowls.

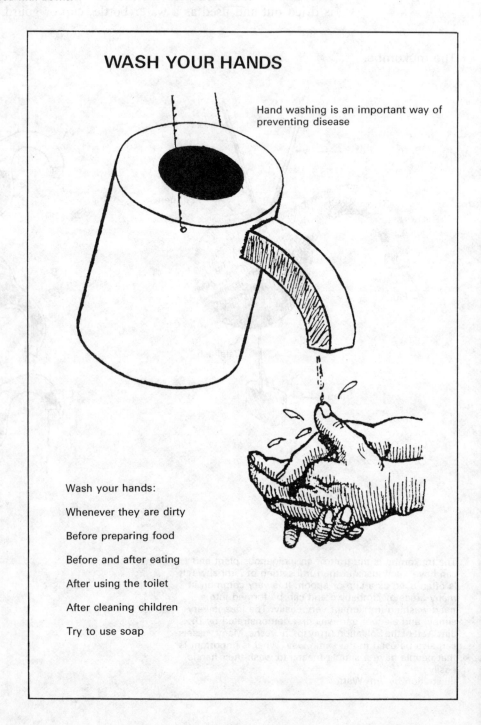

WASH YOUR HANDS

Hand washing is an important way of preventing disease

Wash your hands:

Whenever they are dirty

Before preparing food

Before and after eating

After using the toilet

After cleaning children

Try to use soap

It can be made simply by taking a suitably shaped fruit, hollowing it out and making four holes in the appropriate places. These are illustrated in the diagram. The largest hole is made through the top of the 'mukombe', so that water can be introduced, two further holes are drilled on either side of the centre line, as shown, for the twine or string which will support the 'mukombe'. Two methods are available to make the outlet hole through which water will pass for handwashing. The diagram shows how, in one method, the end of the neck of the 'mukombe' is sawn off a little past the bend in the neck. A 'bush cork' is now made so that it fits in the open neck of the 'mukombe'. The cork can be made of wood from a tree branch. A notch or slot is cut into the cork so that when it is fitted into the neck of the 'mukombe' some water can pass by for handwashing. The notch is fitted facing downwards. A simpler second technique involves drilling a small hole at the end of the 'mukombe' neck, facing downwards.

The twine or string is threaded through the two holes made for it and these are attached to a horizontal stick, as shown, and the two strings are tied to a supporting structure like a tree or the latrine. When suspended the 'mukombe' should hang as shown in the diagram.

Once it is hung, water is added to the 'mukombe' so that it is nearly full. The hand is now used to tip the gourd so that water passes up the neck. When the hand is released, the 'mukombe' will return to the resting position, but some water will be retained inside the end of the neck, as shown. This water will drain through the slot or hole in the end of the neck and is sufficient to wash the hands. More water can be supplied by tipping the 'mukombe' again. As an additional hygiene measure a bar of soap can be attached to the 'mukombe'.

The 'mukombe' can be fabricated in PVC, fibreglass and also tin sheet. It is the model made by the tinsmith which appears to be the promising man-made version of the hand-washing device. It has been tested by a research team at the Blair Laboratory, and is effective for washing hands, even without soap. Most of the test group who were given the mukombe used it regularly, with the result that their hands, when tested, were consistently cleaner than another group using more traditional hand-washing techniques.

Drinking water quality for rural areas

Many standards have been laid down for the quality of drinking water for rural areas, and these vary from one source to another. Similarly much stricter standards have been laid down for urban water supplies, where one central source may serve millions of persons. Clearly in such a case the standards must be strictly adhered to.

In the rural areas however, water is derived from many different sources and almost all of these are untreated with disinfectants. It is known however, that water of relatively high quality can be derived from the ground, if the standards of sanitary protection of the source are good enough.

In practical terms, it would be impossible to condemn every possible source of water that fell in quality below a certain standard, unless this could be achieved in a practical way such as relocating a well that had been dug too close to a source of contamination such as a latrine. Furthermore it is practically impossible at present to consider taking samples of water from all the potential sources that people drink from, simply because this would amount to tens of thousands of samples. The quality of water extracted from the ground also varies with season. The lowest counts for faecal *E. coli* for instance are found in the winter, in both protected and unprotected groundwater sources, with higher counts being found in the warmer and wetter months.

At the present time, the most practical approach to the problem of improving and maintaining the quality of water delivered in rural water supply schemes is not to impose a set standard, but to insist on adequate measures of sanitary protection which significantly improve the quality of water, compared with traditional sources that might have been taken otherwise. It is both possible and practical at the present time to set certain physical guidelines for the proper and adequate sanitary protection of water sources, especially those which are not piped supplies. When underground sources of water are adequately protected by physical means and sited properly, the standards of water quality improve significantly compared with unprotected sources.

It must also be remembered that even if a perfectly pure, but untreated source of water is drawn from a handpump, the level of contamination will rise immediately it enters the vessel which is used to carry the water to the homestead. Furthermore, the water may be transferred from one vessel to another in which case the level of contamination increases further. To set a strict level for the quality of water delivered under these conditions would be of little practical value.

There is also considerable current dispute as to whether the consumption of water which falls below some existing standard for quality will have any noticeable effect on health in any case. For many years it has been assumed that the physical consumption of water below a certain standard will result in poor health. However, there is very little scientific evidence which shows conclusively that the provision of improved drinking water alone will result in an improved health pattern in the population. Whilst improved drinking water sources are vital, since they overcome the harmful effects of gross pollution, which can occur in unprotected sources, the full effects of improved drinking water quality can only be gained when individuals practise improved personal hygiene. Some studies show clearly that the provision of soap together with a campaign to improve hand-washing practice had a far greater impact on diarrhoeal disease than the simple provision of improved water supplies alone. The importance of good hygienic practice as a meaningful method of reducing morbidity due to enteric disease is becoming more widely acknowledged.

These clear facts pose certain questions when strict standards are set for the quality of water in rural areas. The important question is — how meaningful are the standards in practice.

According to the International Reference Centre for Community Water Supply and Sanitation (The Hague, Netherlands), the levels of Coliforms present in acceptable drinking water should be less than 10 per 100 ml sample, and the number of faecal *E. coli* should be less than 2.5 per 100 ml sample.

Further guidelines for the quality of drinking water standards for community water supplies are presented from the same source, below:

Table 8.

Water quality parameter	measured as	Highest desirable level	Maximum permissible level
Total dissolved solids	mg/l	500	2000
Turbidity	FTU	5	25
Colour	mg Pt/l	5	50
Iron	mg Fe^+/l	0.1	1.0
Manganese	mg Mn^{++}/l	0.05	0.5
Nitrate	mg NO_3^-/l	50	100
Nitrite	mg N/l	1	2
Sulphate	mg SO_4^{--}/l	200	400
Fluoride	mg F^-/l	1.0	2.0
Sodium	mg Na^+/l	120	400
Arsenic	mg As^+/l	0.05	0.1
Chromium (hexavalent)	mg Cr^{6+}/l	0.05	0.1
Cyanide (free)	mg CN^-/l	0.1	0.2
Lead	mg Pb/l	0.05	0.10
Mercury	mg Hg/l	0.001	0.005
Cadmium	mg Cd/l	0.005	0.010

It is emphasised however, that these water quality guidelines should always be applied with common sense, particularly for small community and rural water supplies, where the choice of source and opportunities for treatment are limited. The criteria should not in themselves be the basis for rejection of a groundwater source, especially for small supplies which are frequently provided from wells, tubewells, boreholes or springs. Under these conditions, it is acknowledged, any standards, including the ones outlined above, cannot be taken too strictly. Such standards are impossible to implement in most rural conditions, and therefore may have little meaning, in practice.

Where the water derived from a source such as a well or tubewell or even a spring or borehole is found to be consistently well below a certain standard, when the counts for *E. coli* may number several hundreds or even above one thousand colonies per 100 ml sample, then this means the well is inadequately protected and requires closer inspection.

Water quality and the sanitary survey

This is perhaps the most meaningful method of ensuring an overall improvement in rural water supplies, especially those derived from boreholes, wells, tubewells and springs.

The sanitary survey is an on-site inspection of the water supply system and particular attention is paid to aspects of the system which may be a source of contamination. No bacteriological or chemical examination of the water derived from a water source can take the place of a complete and thorough knowledge of the conditions at the site of the supply. Of particular importance are inspections of the following aspects of the supply.

1. The location of the water point, and whether this is on raised ground, and at a distance from lower ground which may become swampy, and more prone to contamination.
2. Depth of water. Generally, groundwater which is very close to the surface is subject to greater chances of contamination from the surrounding areas than water which lies more deeply in the ground, where the effects of filtration by the soil are greater.
3. Distance from potential sources of contamination such as:
   a) Latrines
   b) Cattle pens
   c) Refuse pits and hollows in the ground.
   This distance should be 30 metres or more
4. The sanitary protection of the lining or casing of the well, tubewell or borehole. The lining should be complete and undamaged.
5. The state of the slab and/or apron covering the water point, whether this is complete or is cracked and whether the water is allowed to run off into an adequate water disposal system.
6. The pumping or water raising system is well sealed (in the case of a handpump) so that water does not run past the pump head back into the well or tubewell.
7. The quality and length of the water run-off system. This is related

to the presence of stagnant pools in the area, which reflect poor run-off. It should be at least 6 metres long and preferably 10 metres.

8. The type of waste water disposal system, whether this is adequate and well away from the water point.

The ideal water point

The ideal water point, such as a tubewell, should be sited on elevated ground, and away from areas where waste water accumulates. It should be well away from latrines, cattle kraals and hollows in the ground. The casing (or well lining) should be well grouted, and the cover slab strong and complete, without cracks, and sloped so that water runs to waste down a long well-sealed water run-off and enters a shallow soakaway or seepage area.

Under these ideal conditions the source of the water is unlikely to be contaminated to any extent, either from underground, or from the surface.

The relationship between water quality and pumping system

The quality of water expected from a well or tubewell will vary according to the type of water raising system employed. Water raised from a well with a handpump can be expected to contain fewer bacteria than a well fitted with a bucket and windlass. However the bucket and windlass is less likely to malfunction, and is cheaper to maintain than a handpump, which may go out of action if an effective maintenance and repair system is not operating. If a handpump fails, people may be forced to take their water from an unprotected source, which will provide water of a lower standard than a well fitted with a bucket and windlass.

For these reasons a compromise must be achieved between the desired water quality and the maintainability of the system. It could be strongly argued that water points which deliver lower quality water, but more reliably are of greater value in the long term than water points which deliver higher quality water less reliably.

However, the desired quality of water might also vary according to the number of users taking water from a particular source. At the beginning of this chapter it was explained that the very strictest standards must be maintained for piped urban supplies, where many thousands of people, and even up to one million persons, may take their water from a single source. The highest standards must be maintained for all piped chlorinated water supplies which serve even a thousand persons from a single source. However, the same standards may not be maintainable with remote handpumped supplies.

Fortunately the highest quality water derived from handpumps is delivered by the Bush Pump, which is normally used by the largest number of people (between 100 and 500 persons). Slightly lower quality water can be expected from a Blair Pump, where the upper hygienic seal may wear, but these are normally used by less people. Bucket Pumps deliver water of a slightly lower quality compared with Bush and Blair Pumps, but are used by less people (up to 60 persons). The water taken from upgraded wells will normally contain larger numbers of bacteria than water taken from a Bucket Pump, but this too will be used by fewer people (4–30 persons) compared with the Bucket Pump. Thus there is a direct relationship between the expected number of users and the type of pump used, and also the expected water quality and the type of pump (or water raising system) used. In practice pumps which are recommended for use by more people are those which deliver higher quality water.

There is also an inverse relationship between 'pump' type and ease of maintenance. The hardest and most costly pumps to maintain (Bush Pumps) deliver the best water, whereas open water holes, which require no maintenance deliver very poor quality water. In between these two extremes lies a series of options — Blair Pump, Bucket Pump, Upgraded well. In each case the ease and cost of maintenance is inversely related to the expected water quality. Upgraded wells are easier to maintain than Bucket Pumps, but in most cases deliver water of lower quality. Bucket Pumps are easier to maintain than Blair Pumps but in most cases deliver water of lower quality. However water delivered from Bucket Pumps and Upgraded wells is certainly quite adequate at the setting recommended for them. In some cases Bucket Pumps or Upgraded wells may actually yield water of a higher quality compared with a handpump, depending on the setting and state of the pump.

Normally Upgraded wells are used by families or extended families and Bucket Pumps used by extended families or small communities. Bush Pumps are used by large communities. In the family or extended family setting, great care will be taken of an Upgraded well or Bucket Pump, and the number of users will be relatively small. Under such conditions, the users can more easily harmonise with the well or tubewell itself, including the micro-organisms that may pass to and from the well via the user. The risk of heavy pollution is low. Where the number of users is high, as in most heavily used communal water points, the potential for complex cross contamination of an open well is far higher, since many buckets may be used, and the environment of the headworks is congested. Under these conditions the same principles can no longer apply. Fortunately a heavily used water point of this type will be fitted with a Bush Pump which offers considerable protection against cross contamination. The Bucket Pump fitted in a small community setting is placed in an intermediate position, being used by more people than in an extended family setting, but yielding higher quality water than an Upgraded well, especially if it is placed on a tubewell.

Thus in judging whether a sample of water taken from a water point has an acceptable standard or not, consideration must also be taken on the setting and the number of users.

For a handpump supply, it is unlikely that many samples will rise above a count of 50 *E. coli* per 100 ml sample, and that most will lie under 10 *E. coli* per 100 ml sample. Since many surface water supplies, which are commonly used as an alternative can be grossly contaminated showing *E. coli* counts above 1000 per 100 ml sample, a source that is consistently less than 50 *E. coli* per 100 ml sample and normally less than 10 per 100 ml sample must be considered a considerable improvement over an unprotected source and should not be condemned.

Table 9. Bacteriological quality of water taken from wells and handpumps used in the Epworth peri-urban settlement area. Data expressed as mean number of faecal *E. coli* per 100 ml sample and the number of samples.

Samples were taken during the period January to November 1988 during a heavy rainy season. Data analysed by the Blair Research Laboratory, Harare, using a membrane filtration technique and membrane lauryl sulphate as a nutrient medium.

Source	Mean *E. coli*/100 ml sample	Number of samples
Poorly protected well	266.42	233
Upgraded wells	65.94	234
Bucket Pump (overall)	33.72	338
Blair Pump (tubewells)	26.09	248
Bush Pump (tubewells)	6.27	281

Explanatory notes

Most pumps were fitted to hand drilled tubewells, other than a few Bucket Pumps fitted to wide diameter wells. The quality of water is related to many factors including the type of water lifting device, the depth of the water in the well, the type of well, the season, and the quality of the headworks.

Bush Pumps provide the best water partly because they draw water from the deepest aquifers and partly because they offer the best hygienic seal at the well head.

Blair Pumps take water from shallower aquifers which are more susceptible to seasonal variation in water quality, this results from ground water movement near the surface during rainy weather (see Groundwater chapter), the leakiness of earlier model Blair Pump heads when worn, may also influence water quality in this category, pump design has subsequently been improved.

Bucket Pump water quality is partly dependant on the type of well and the type of installation. The water is best when the Bucket Pump is fitted to a tubewell, when the flushing effect is most pronounced.

The results also show a pronounced difference between poorly protected wells and Upgraded wells, the technique of improving the headworks of traditional wells, as described in the Upgraded Well chapter, reduces the number of bacteria in the water by a factor of about four times, this is a significant improvement, and has led to the introduction of pilot experiments where family and extended family subsidies are being tried for improving existing shallow wells.

These are research results in which some pumps and wells were placed in sites which were not necessarily ideal, the results do not reflect the best possible figures for faecal *E. coli* that might have been possible from any particular pump source. This study and earlier studies have been used to determine the nature of water quality derived from different sources and different pumping systems.

Thus if the standard of an *E. coli* test is to be taken as a guide to drinking water quality, then the following tentative guidelines for rural water supplies, using handpumps and Bucket Pumps, are suggested.

1. Supply found with 1–10 *E. coli* per 100 ml sample — supply satisfactory.
2. Supply found with 11–50 *E. coli* per 100 ml sample — supply needs further testing.
3. Supply found with 51 *E. coli* or above — supply needs investigation and if repeatedly above this value requires improving.

These figures are based on research work undertaken by Blair Research Laboratory in the Epworth area near to Harare. The bacteriological quality of water derived from Bush Pumps, Blair Pumps, Bucket Pumps, Upgraded wells and poorly protected wells was analysed for an eleven month period (January 1988 — November 1988). The mean *E. coli* per 100 ml sample for the type of facility is shown together with the number of samples analysed. Explanatory notes are provided beneath the figures.

Nitrate

The level of nitrate should also be considered. According to the IRC report referred to earlier, the highest desirable maximum level for nitrate is 50 mg/litre and the maximum permissible level is twice this. It is known that nitrate concentrations over 45 mg/litre in drinking water are potentially hazardous to health. Current practice in urban situations is to condemn any water with a nitrate concentration above 10 mg/litre. High nitrate levels in water usually indicate the presence of organic pollution in the ground which can originate from fertilisers, sewage discharge from tanks or from on-site sanitation like pit latrines. An upper limit of 50 mg/litre for rural water supplies should be set, with a desirable maximum of 30 mg/litre.

Clearly water samples should be odourless and colourless and preferably have a pH value between 6.5 and 9.2. Dissolved salts should not exceed 1500 mg/litre.

It should be pointed out that all the guidelines outlined above apply to untreated water supplies derived from wells, tubewells, boreholes and springs, and that where a piped supply is treated then higher standards must apply, especially for bacteriological standards.

Any untreated water contains active plant and animal life and bacteria. Most of these bacteria will not be harmful, and indeed many are very valuable to life processes. It is the pathogenic faecal bacteria which pose the greatest health hazard, and these can be detected by the presence of *Escherishia coli* (*E. coli*) referred to earlier.

Each Provincial Office of the Ministry of Health is now equipped with a simple bacteriological testing kit (DelAgua field testing kit). This apparatus is used to test for *E. coli* and other parameters for water quality. This single kit can test up to sixteen samples at one time.

However, until more water-testing apparatus becomes available in the rural setting, it will be necessary to rely largely on sanitary surveys made on existing water supplies, and the adherence to sanitary measures which improve the potential quality of supplies under construction. The main aim of health workers should be to improve existing supplies as well as they can using techniques which are available, and not to condemn supplies because a single test shows the presence of *E. coli*. The latter case may prompt people or even force people to abandon improved but lightly contaminated sources of water in favour of the only alternative, which may be unimproved and heavily polluted.

If a process of upgrading and improving existing traditional sources is maintained then it is unlikely that public health will be impaired, and will almost certainly be improved. The aim of improving what was there before, must be the aim of every health worker.

The purification of water

Water taken from potentially contaminated sources like rivers, dams and pools should be purified before it is consumed. If no other means are available, the water should be boiled for several minutes to ensure the bacteria are killed off.

Solar radiation also kills bacteria. If a clear glass or plastic bottle is filled with relatively clear water from a well or river, capped and turned on its side and left in the strong sunlight for a period of 4 hours, most of the bacteria will be killed.

Similarly low strength chlorine solutions like '*Jik*' or '*Javel*', manufactured for bleaching can be used to purify water. A teaspoon of '*Jik*' in a 10 litre bucket disinfects the water in 60 minutes, and with two teaspoons, disinfects the water in 30 minutes. This is a low cost technique costing about one Zimbabwe cent (½ US cent) per litre to disinfect. (Data of Civil Engineering Department, University of Zimbabwe).

Sand can also be used to purify water. This a natural process and one that forms part of the system in many large scale purification works. It also works on a small scale.

When water passes through the soil, it also becomes purified, as pathogenic bacteria do not find either sand or the soil a good medium in which to multiply. They tend to die off. This explains why water taken from adequately protected wells and tubewells excavated in the soil yield water with very few bacteria in it.

This section of the book deals with methods of purifying water, and describes the use of sand filters and the small scale treatment of water with chlorine.

A knowledge of how to purify water taken from unprotected sources is important.

Sand filtration of water

Sand has been used to purify water for over a thousand years, and it still remains one of the most dependable methods of making water fit for drinking. Water taken from sandy river beds is generally pure, because it has percolated through the sand grains where the bacteria harmful to man die out. The process is one that occurs in nature and therefore if the conditions are right, it works very satisfactorily.

How sand purifies water

In nature, pathogenic bacteria that cause disease, thrive well where nutrients are plentiful and the temperature is warm. When placed in a foreign environment like sand, however, the pathogens die out. Sand offers a natural and effective method of purifying water.

The effects of sand on water are complex and make use of several mechanisms. Many bacteria are consumed by the skin on the surface of the sand known as the '*schmutzdecke*' made of algae, diatoms, protozoa and other organisms. Water then passes between the sand grains and a process of adsorption takes place on this surface by electrical and chemical bonding and mass attraction. Each cubic metre of sand provides 15,000 m^2 of surface area for the adsorption of bacteria to take place. In addition, the pores or open spaces between the grains occupy 40% by volume of the sand. Water flowing through these spaces slows down and sediments out. As the depth from the surface increases, the quantity of organic matter decreases and the struggle between the various organisms becomes fiercer. The pathogenic bacteria cannot compete with other organisms more suitably adapted to these conditions. In such an environment disease carrying bacteria perish.

This remarkable process takes place in all sandy river beds, and is used widely throughout the world as a water purification process, even in major cities, together with disinfection with chlorine. The water supply of Amsterdam in Holland is purified simply by passing it through the sand dunes on the coast. Where water is abstracted from sandy river beds by pumping, it has great purity.

Sand filters can be made in many sizes, from a family unit contained in a 200 litre drum or cement water jar, to very large units designed for small communities. Two units described in this chapter, one for the family and one for a small community of about 300 persons.

Sand filters will normally be used where the source of water is known to be polluted. This will normally be water pumped from a dam or river where bacteria counts can be high. Whilst sand removes bacteria effectively from the water, it has difficulty in coping with large volumes of sediment which may be present in the water. Ideally water should be added to a filter when it is mechanically clean, and this may involve pre-treatment in the form of sedimentation, or passing through a gravel filter first. If the water contains a lot of sediment and this is not removed, the surface of the filter will become clogged rapidly.

The family sand filter

The 200 litre drum is a convenient container in which to construct a family sand filter. However a brick built container can also be used, or even traditional pots or cement jars can be used. The use of bricks is probably the best and most convenient.

Sand filters can be contained in many vessels including traditional pots.

Cement jars can also be used to contain sand filters.

257

The design of the working components of the filter using a 200 litre drum should provide at least 50 litres of water per day, enough for drinking and cooking water for a small family of four or five persons. Larger families will require larger filters which hold more sand. These should be made from bricks and mortar.

The description below applies to a drum, but the same technique can be used in a brick built container. The drum is cleaned out and a hole is made two thirds of the way up so that an outlet pipe can be fitted. If a 12 mm tap is fitted as a means of taking water from the filter, the hole should be large enough to take a 12 mm galvanised barrel nipple. The nipple is cleaned and fitted through the hole and is either secured by brazing or with hard-setting putty. In a brick filter a 12 mm pipe is introduced and cemented within the wall.

The inside of the drum is then painted with a protective covering such as bitumastic paint, which is allowed to dry thoroughly. This will not be necessary in a plaster lined brick built tank.

Before the sand is added it is necessary to fit a water collecting pipe (underdrain system) at the bottom of the drum and cover this with washed gravel. The water-collecting pipe can be made with a ring of 20 mm plastic hosepipe about 350 mm in diameter. This is connected to the outlet pipe already mentioned by a short length of hosepiping. A number of saw cuts or drilled holes are made in the hosepipe ring, and this is laid on the bottom of the drum or tank so that the saw cuts or holes face downwards. Once the pipe connection has been made between the ring and the outlet pipe, a layer of well-washed gravel about 75 mm deep is laid over the bottom of the drum or tank, and levelled out.

It is now necessary to find some good, sharp, very clean river sand to add to the drum or tank. It is essential that the sand is of a good quality, preferably taken from a river bed and thoroughly washed. This is added to the drum so that it is about 500 mm deep. The level of the sand should come up to just below the level of the outlet pipe. This arrangement is made so that even if the tap is left on, and water drains out of the filter, a small layer of water remains above the sand. The sand must never be allowed to dry out, otherwise the biologically active ingredients in the sand, which are important to the purification process, will die out.

Some form of lid is made to cover the drum, and this is made with a hole in it through which water can be poured. In order to avoid erosion of the sand caused by the water which is being added, a flat stone is laid above the sand, directly underneath the hole in the lid. This flat stone is best raised above the sand by means of three or four smaller stones beneath it.

When the filter has been completed, it must be thoroughly flushed through with clean water. Once this is complete a daily routine of adding water can be maintained.

Freshly washed river sand will take a few weeks to ripen and develop biologically active ingredients which help purify the water. An improvement in taste and colour should be noticed almost immediately.

Since the water poured into the filter may be slightly turbid at times, the filter may, after some weeks or months, begin to clog up and the delivery rate slow down. When the delivery rate is too slow for con-

Section through family sand filter made from a 200 litre drum. The same technique can be adapted in many ways with local materials and methods. The drum can be replaced with a brick built tank and the tap with a wooden peg fitted into the outlet hole. A short length of piping will be required for the under-drain system.

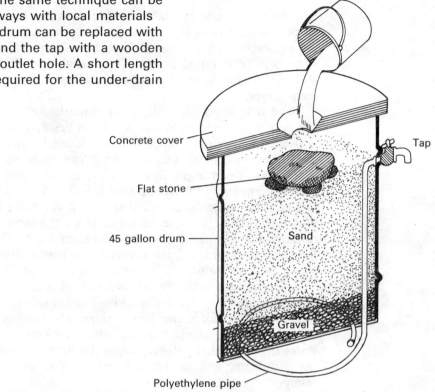

venience, several inches of sand should be removed, thoroughly washed and replaced in the filter. If the source water is too muddy, it must be allowed to settle in a drum or tank before adding it to the filter.

When used in a family situation, a routine of adding water to the filter must be developed. The drum should be kept topped up regularly with water so that the reservoir section of the drum or tank is kept relatively full. This accounts for about one third of the total volume of the tank. Ideally the tank should be topped up several times a day, and water should be withdrawn through the tap in small quantities only — for drinking in a cup, for instance. The flow of water through the filter should be kept as slow as is conveniently possible. The water can be kept trickling into a clean container for a longer period.

The filter must not be drained completely in one go to fill a bucket, and then left standing before more water is added to the top. This will cause surges of water through the sand, the retention time of the water in the sand will not be long enough, and water quality may not be improved. By careful addition and draining of the filter, water quality can be improved significantly.

Larger containers can be built with bricks and mortar in the same way. As a general rule, add 75 mm gravel to the underdrain pipe and then fill with washed river sand two thirds of the way up the filter. Have the outlet pipe slightly above the sand level. The upper third of the tank holds water. Filters should always be covered with lids.

Making a small community sand filter

The unit described below is capable of supplying enough water for a small community of up to 300 persons, assuming a consumption rate per person of no more than 20 litres of water per day. It is assumed that water will be pumped from a river, canal or dam into a reservoir first and led via the filter to a storage tank or to water points within the community.

Since it is important that water runs through the sand filter slowly and continuously for it to function at its best, a relatively large reservoir is needed and also a storage tank for filtered water to collect in. Since sand filters operate at their best when the water which is fed into them is relatively clean, some form of pre-treatment process is necessary, to remove as much of the sediment as possible. If a reservoir already exists this can be used, and modified so that it is more effective as a settling tank. Alternatively, water can run from the storage tank through an upward flow gravel filter to the sand filter itself. The main function of the settlement tank and the upward flow gravel filter is to trap as much silt as possible before the water enters the sand.

If the water is being withdrawn from a dam, river or stream with a pump, it may be easier to dig an infiltration gallery, which is basically a well, near the bank of the dam or river. The method can be the same as described in the chapter on the construction of a well. In this case water in the ground will seep from the dam or river into the well, and the sediments will be filtered out by the soil before the water is pumped.

The infiltration gallery

If the banks of the river or dam from which the water is taken are built up in soil and not rock, a well or infiltration gallery can be excavated close to the source of the water but above the highest water mark. This method can serve well for removing sediments from water derived from a river or dam before they are pumped. When water is extracted in this way, it may be pure enough to be piped without sand filtration or chlorination. However it is a good method of improving the quality of water and preparing it for sand filtration.

The well can be made 1.5 metres in diameter and lined with bricks or concrete rings as described in the well chapter. If larger amounts of water are required, the surface area of the walls of the well may need to be larger. In this case a wide well can be dug, perhaps two or three metres in diameter about 10 metres away from the high water mark, and as deep as possible. Concrete rings can be laid centrally within the 'gallery' and backfilled with well-washed river sand. After three metres of sand backfill, a layer of thick plastic sheeting can be laid in the annular space and the remaining depth filled with soil. The rings and backfill are brought to the surface, and a concrete cover slab and apron built. A suitable pump can be fitted to the gallery/well and water piped to a header tank.

The reservoir settling tank

A large reservoir made of bricks serves this purpose well. In order for the settlement tank to operate effectively it is necessary to avoid turbulence caused by water being pumped into the tank. One way of doing this is not to pipe incoming water, to the top of the tank, but to lead water through a wide pipe to the bottom of the tank. This pipe can be made of PVC (200 to 300 mm diameter) or more cheaply from bricks. One method involves building a brick chimney in the middle of the reservoir (6 bricks per course), with a 225 mm × 225 mm inside cross-section. The base of the chimney is built slightly wider so that an opening can be left at the base. Water from the pump is directed down the chimney and enters the tank from the bottom. An arrangement of stones around the base of the chimney reduces the disturbance further. When water is led to the bottom of the tank in this way, the most turbid water (and hence that retaining most bacteria) remains nearer the bottom with clear water appearing at the surface.

Apart from being less turbid, surface water becomes purer because sunlight (ultra violet light) kills off pathogenic bacteria. Thus it is very desirable that the water which is fed from the settlement tank to the filter is taken from the surface. One effective method of doing this in a tank with varying water levels is to build a water collector that floats near the surface. This has been successfully achieved by making a float of 40 mm polypipe in the form of a ring, and wiring on to this another ring fitted with a tee piece. In the lower water collecting ring a series of holes are drilled. Most of these are 10 mm in diameter and are drilled on the underside of the ring. A small number of 5 mm holes are drilled on to the upper surface to allow air to escape when the ring is filling with water. Water which enters the ring is led through a tee piece to an outlet pipe which passes through the lower wall of the tank to the filter. By arranging the plumbing of the tank in this way, turbulence is reduced and surface water is always collected. These two features in themselves will improve the quality of the water and make it more suitable for sand filtration. If a settlement tank is being built in brick, this can be built on raised ground about 4 metres in diameter and 2 metres high, and left uncovered.

Upward flow gravel filter

This is an alternative to the settlement tank and is more effective at reducing silt levels, although it will require backwashing at times. If the storage tank which receives water from the pump is small, it is worthwhile to build an upward flow gravel filter to eliminate most of the suspended solids in the water before it enters the sand filter. In this technique water is led from the source into the bottom of a brick-lined tank. Water passes through the gravel layers and the silts settle out on the gravel. A slotted pipe is placed inside the gravel so that it taps off water just beneath the surface of the gravel. Water from this pipe is then directed into the inlet of the sand filter.

An upward flow gravel filter suitable for the sand filter described later should be about 1.5 metres wide and 1.5 metres deep and can be built up with brickwork in the ground provided that the sand filter lies below this level. Normally the system is built on a slope so that water can gravitate from the upward flow gravel filter into the sand filter and then finally into the storage tank or standpost system.

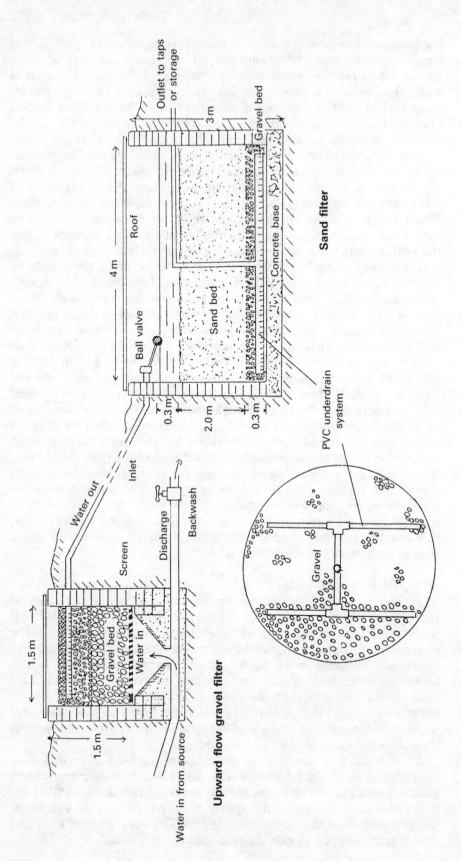

Section through small community gravel and sand filter.

Since the gravel filter will require backwashing from time to time, a system like the one illustrated should be built. Water from the source is led through a smaller pipe (40 mm PVC or 25 mm GI pipe) to a wider 75 mm tee. One arm of the tee which faces upwards allows water to pass into the chamber of the upward flow gravel filter, the other arm directs water into a wide (50 mm or 75 mm PVC or 50 mm polypipe) backwash pipe. This larger pipe is led off to a lower level and is fitted with a 50 mm gate valve. Under normal operation this gate valve is closed and the water is directed upwards through the gravel. The chamber of the gravel filter is designed so that water passes upwards through a cone shaped base through a heavy duty steel screen and then through layers of gravel held up by the screen. The screen itself can sit on a circular shelf of mortared brickwork laid as shown in the diagram. The cone shaped section at the base of the filter can be built up in concrete.

All the walls of the brick-lined tank are plastered with cement mortar to make them watertight. The gravel filter outlet pipe can be made of heavy duty PVC pipe, 50 mm class 16 being a good size. This pipe is slotted on the underside and cemented through the brick wall of the gravel filter tank.

When the cement work of the filter is cured, the strong steel screen is laid on the shelf at the base of the filter and a 0.5 metre deep layer of 25 mm granite chips is laid on top of this. An upper layer of finer gravel or 6 mm granite chips is laid above the lower layer, also 0.5 metres deep. The upper layer should cover the outlet pipe. Finally a cover is made for the gravel filter from corrugated iron sheets or ferrocement.

Under normal operating conditions, water from the source which will normally be a tank is led through the inlet pipe, with the backwash valve closed. Water will pass upwards through the gravel and pass into the water outlet pipe, and then into the sand filter itself.

The sand filter

Water led from the infiltration gallery, settlement tank or upward flow gravel filter enters the sand filter through a ball valve. A convenient way to build the filter is to construct a brick tank on a concrete base — with most of the structure lying below ground level. A tank suitable for a small community should hold about 36 cubic metres of water, and a convenient size is 4 metres in diameter and 3 metres high. A hole is dug in the ground about 2 metres deep and just over 4 metres across. A concrete foundation is laid in the bottom and a brick wall (110 mm thick) is built up to about 1 metre above ground level. The actual arrangement will depend on local conditions. Tanks can be built entirely below ground level or entirely above it, depending on the lie of the land. The tank is plastered internally so that it is watertight. An inlet pipe and an outlet pipe are inserted into the wall during construction. The inlet pipe can be made of 25 mm GI pipe to allow attachment to a 25 mm low pressure ball valve. This is fitted near to the top of the tank (see diagram). The outlet pipe is made of thick walled 50 mm PVC pipe and a section of this is fitted through the brick wall approximately 500 mm below the top of the tank.

The water collection pipes (under drain system) are made of 50 mm class 16 PVC pipes. One convenient method of doing this uses PVC tee pieces (see diagram). Water enters the PVC pipe through a series of saw cuts made in the lower half of the pipe. A vertical pipe leads water away from the collector pipes to the outlet pipe through a PVC 90 degree bend.

A volume of 6 mm granite chips is then thoroughly washed and laid over the pipes to a depth of 300 mm. This will require about 4 m^3 of granite chips. A volume of thoroughly washed sharp river sand is then added to the filter so that the final depth after flooding amounts to 2 metres. This is about 24 m^3 of sand. The best sand is taken from a river bed (sharp sand) and washed so that all organic matter is removed.

Once the sand has been added, water can be passed through the system to allow the sand to settle. This water should be allowed to run to waste for about one week, to flush out sediments and waste matter. An initial dose of chlorine will also help to sterilise the sand bed. It is preferable to arrange the levels of the sand so that they lie just below the outlet pipe. By doing this, the sand will not dry out, even if the water supply to the filter is cut off. When fully flooded, approximately 30 to 40 cm of water should lie above the sand.

An ideal flow rate for the filter of this size is 300 litres of water per hour or approximately 5 litres per minute. When the levels of the inlet and outlet are set a gate valve fitted to the outlet pipe can be adjusted to allow water to pass at the appropriate rate. The sand filter can be designed smaller if less people are served. 12 m^3 of sand can filter enough water for 150 persons. A roof should be fitted over the sand filter. This can be conveniently made of corrugated iron sheets, supported by a central gum pole.

Water storage tank

Since sand filters work at their best when water is allowed to pass through them continuously, a storage tank is very desirable in the system. Obviously this must be built below the level of the filter so that water can gravitate into it. The level of water in the storage tank is regulated by a ball valve. The storage tank should be built so that it holds about 2 days supply (in the system described this will be approximately 12 m^3). If the flow of water through the filtration system is very reliable the storage tank capacity can be lowered. A brick-built tank serves well as a storage tank, and this should be covered to prevent further contamination. A removable cover should be located in the roof to check on the ball valve. A 12 m^3 tank is approximately 2 metres deep and 3 metres in diameter. If the storage tank is elevated well above the community living area, water can be piped from the storage tank to a series of taps in the living area. This is described in the chapter on water point design.

Maintenance

The biological layers near the surface of the sand become very active and important to the purification process. If they are disturbed the efficiency of the filter is temporarily reduced. When a reduction in the flow rate is noticed due to clogging of the upper levels in the sand, the water above the sand should be drawn off and a layer of about 150 mm of sand removed and washed thoroughly and replaced. The flow rate will be restored.

The slower the water runs through the sand filter, the higher will be the quality of the water at the outlet. If the system is required to provide water for more people at a later stage, a second filter can be built in parallel with the first.

Backwashing the gravel filter

When the silt loading on the gravel filter becomes too high the flow of water into the sand filter will be reduced. This system now requires backwashing. This process entails opening the large gate valve fitted to the discharge pipe, and throwing fresh water into the top of the gravel bed from above. If the source of the water is a pump or is drawn from a gravity source, then the supply can be diverted so that the water is directed on top of the gravel. The silt deposits collected on the gravel will then be washed down through the screen, down the conical outlet and into the water discharge pipe to waste. Once the water is clarified and the discharge pipe valve is closed, the water flows in the normal direction, upwards through the gravel and into the sand filter.

The system described above, with a upward flow gravel filter and a $24 m^3$ sand filter were built and tested by the Blair Research Laboratory at Henderson Research Station in 1977. Many bacteriological tests have been carried out on the water delivered by these filters. They have continued to provide potable water for a period of over ten years with a limited amount of maintenance.

Water treatment with chlorine

Many of the diseases which are common in the communal lands are carried by water especially from unprotected wells and water holes, rivers and dams. Dysenteries, diarrhoeas and typhoids can arise as a result of drinking water which is infected. Unless water is taken from a protected source like a borehole, a protected well or a sand filter it is likely to be contaminated, and some form of purification is essential. The disease-carrying organisms found in water can be effectively killed by disinfecting the water with chlorine. In whatever form the chlorine is used (alginate, HTH, chloride of lime) the dosage of available chlorine must be sufficient to effect complete purification. The process is an oxidation; the chlorine combines with the hydrogen of water, liberating free oxygen which rapidly oxidises matter such as bacteria present in the water. The amount of oxidisable matter present in the water is important, since much of the chlorine may be used up in oxidisation of organic matter other than bacteria. Properly sedimented and filtered water therefore requires less chlorine to sterilise it compared with raw unclear water. Alginate is the most useful source of chlorine since it is fairly stable in storage. Chloride of lime is cheaper but less stable and tends to lose its chlorine when stored poorly.

Water delivered to small-scale treatment plants is usually pumped from a dam or river. Where water is taken from a borehole or protected well it is already filtered by the ground and is generally pure. However, water can become polluted in storage tanks and reservoirs, and it is not always possible to guarantee water purity. The addition of chlorine to water will guarantee purity.

Two types of water purification systems have been used by the Ministry of Energy, Water Resources and Development. These are the single tank system and the twin tank system.

The single tank system

In the single tank system water is pumped into a 22,500 litre tank (5000 gallons) so that it is full. A single 50 g cup of alginate is added to the water if it is relatively clear. If the water contains some organic matter it is wise to place 1.5 to 2 cups (100 g) of alginate in the water. The alginate is mixed in a bucket of water and thrown into the tank when it is filling to ensure thorough mixing. The amount of free chlorine left in the water should be at least 0.3 ppm (0.3 mg/litre). This should be allowed to react with the water for at least 30 minutes. If there is some doubt about the organic content of the water, a chlorine level of 0.5 ppm should be maintained. Water from such a tank is normally passed through a rapid sand filter (swimming pool filter) before it enters the domestic supply. The filter is backwashed periodically — the frequency depending on the physical quality of the water. The presence

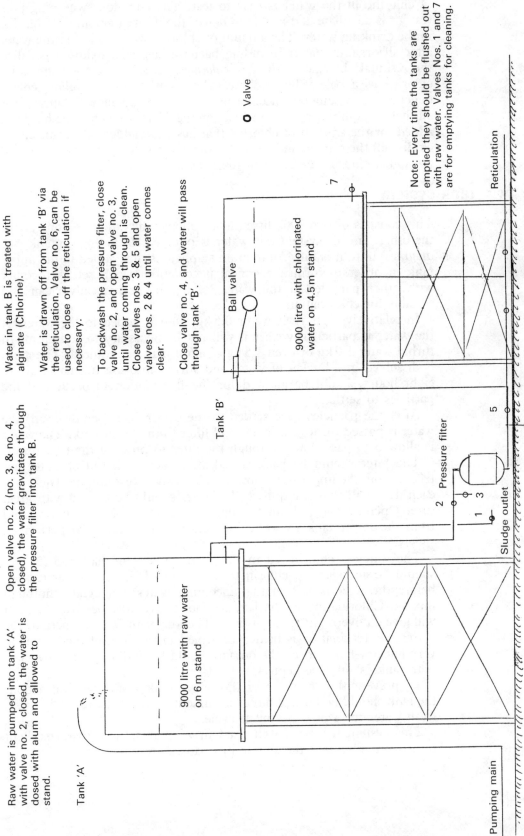

Operating procedure for twin tank system.

of chlorine in the water is easy to test. The common swimming pool test kit is adequate and is able to detect the desired amount of chlorine in the drinking water. The amount of chlorine which is left in the water after all organic matter including bacteria has been oxidised is called the residual chlorine or the *free chlorine residual*. A test kit shows up chlorine as a pale yellow colour. The intensity of the yellow colour depends on the amount of chlorine — but a pale yellow colouration is all that is required. No colour reaction means that no chlorine has been added, or the amount of chlorine that has been added is insufficient to oxidise all the organic matter and provide a residual. It is better to add too much chlorine rather than too little.

The twin tank system

This consists of two 9000 litre (2000 gallon) tanks mounted one above the other. In this system raw water is pumped to the higher tank (see diagram) until it is full. Aluminium sulphate is then added to the upper tank in order to form a chemical flocculation which settles together with other particles in the water leaving it mechanically clean and suitable for chlorination.

In relatively clear river water 0.5 kg alum is added to the water. If the water is particularly clear it will not be necessary to add alum. In turbid water 1.0 kg or even 1.5 kg of alum may be required. The alum is dissolved in a bucket of water and stirred into the full tank. Six to eight hours should be allowed for the flocculation to occur and the particles to settle.

After the particles have settled in the upper tank (see diagram) the water is passed through a rapid sand filter to the lower tank. The water is allowed to pass slowly through the filter to produce the best result.

The filter should be backwashed after each tank full of water has passed from the upper to the lower tank. A short few minutes backwash each day is all that is required. The filter should be treated with great care. Opening valves from the upper to lower tank for filtering and from the lower tank to waste for backwashing should be performed slowly.

The water is chlorinated in the lower tank. One half cup (25 gm) should be sufficient to chlorinate the water. This can be carried out before the tank is full so that incoming water will create thorough mixing. Chlorinating in the late afternoon gives the best results as it will give an overnight reaction time. The longer the contact period, the better. If chlorination is being performed in the heat of the day or in very hot low-lying areas, the dosage should be doubled. The covers of both tanks should be kept closed.

To prevent the corrosion of the upper tank from chemical action, the tank should be flushed out every time it is emptied and refilled with fresh water for a new treatment cycle.

This system has been well tried in Zimbabwe and gives excellent results provided that the simple procedures are followed.

RURAL SANITATION

The Blair Latrine and how it works

The Blair Latrine was developed in Zimbabwe by the Ministry of Health as a means of improving the conventional pit latrine. It is a type of latrine known as a VIP — a ventilated improved pit latrine. Blair Latrines offer a safe and reliable method of excreta disposal especially suitable for rural areas where water is scarce. Blair Latrines, when properly built, are odourless and the fly problem is reduced to a minimum. They are best built as single or double units for the family or as multicompartment units for schools. They do not require water to operate, although water is required for cleaning. Because they are simple, they may be more dependable than many waterborne units and are therefore valuable even in areas where water may be available.

The Blair Latrine was specifically developed to overcome the problems of odour and fly breeding commonly found in pit latrines. Some pit latrines are almost odourless and flyless even without a pipe, but these are rare. They occur when the pit is very deep, say over five metres, when the latrine structure is dark and cool, and when the latrine is used only by a small number of people and kept very clean. Such conditions rarely occur in practice in the communal lands, and most unprotected latrines smell rather badly and are infested with flies and other insects.

When a latrine is odourless and free of insects it is used and appreciated more. Because it is a private place it can also be used as a bathroom.

Early Blair Latrines were fitted with doors and large ventilation pipes. The door system was replaced by a spiral structure which guaranteed semi-darkness at all times.

The Blair Latrine is popular for these two reasons, and since 1980 over 100,000 have been built. It is now the latrine technology of choice for all rural areas in Zimbabwe, and the constructional programme is increasing year by year. This history suggests very strongly that the basic principle behind the design is sound and actually works in practice.

The latrine works well because it employs the forces found in the Natural World to make it operate. Such forces are dependable and can be guaranteed to operate over a wide range of conditions. Two basic forces are operating, one concerned with the movement of air in pipes, the other with the instinctive behaviour of flies.

Pit ventilation

The illustrations on the next pages show the effect of air currents and solar radiation on the flow of air in the vent pipe. Where the pipe is fitted over the latrine slab as shown, any air movement across the top of the pipe will cause an updraught in the pipe. Air is literally sucked out of the pipe by the air passing across the top. The air forced to rise up the pipe is replaced by new air which is sucked in through the squatting hole in the slab. The squatting hole acts like an air inlet with the pipe acting like an exhaust, as in a motor car. In the Blair Latrine air is continuously passing through the squat hole and the vent pipe, often rising in the pipe at a rate of over a metre per second when a good breeze is blowing. When this air movement is taking place, it is impossible for the foul gases in the pit to escape up through the squat hole into the latrine house. All the odours pass up the pipe and are diluted in the atmosphere. The interior of the latrine remains odourless.

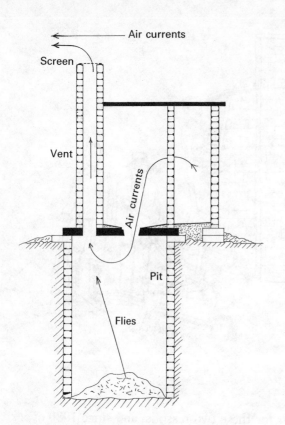

Blair Latrine.

Although wind is normally the main force which draws air through the vent pipe, this is not the only mechanism operating. On hot still days, the pipe ventilates without wind. In this case the sun heats the wall of the pipe, and this in turn heats up the air inside the pipe. Since hot air rises, the air will pass up the pipe and cooler air will be drawn in from the pit. This mechanism works particularly well in thin-walled pipes, which heat up quickly. It is less effective in thick-walled brick pipes. Thin-walled pipes, made of asbestos, steel and PVC are usually coloured black or grey to help this effect.

The effect of the pipe can be demonstrated in a 'smoke test' when a smoky fire is lit inside the pit, usually with a mixture of sticks and grass. Once the flames have settled down, the smoke is drawn up the pipe, with almost none coming out of the squat hole. This test is very valuable when checking for the construction of a Blair Latrine. If there has been an error in the construction, and the pipe does not ventilate well, smoke will come out of the squat hole. It is important to check for possible mistakes in construction. The smoke test is illustrated in the diagram.

The ventilation pipe of a Blair Latrine still functions if there is no structure. A good demonstration of the effect can be made by covering a shallow pit with a concrete slab made with two holes, one fitted with a ventilation pipe. The air will flow in through the open hole, and will pass out of the vent pipe as shown.

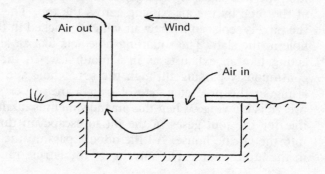

If there is no wind and the sun is shining on the pipe, the air still rises in the pipe.

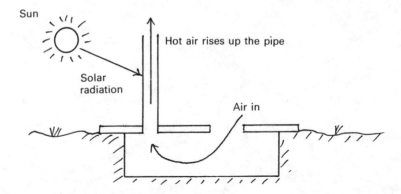

These effects can be demonstrated with the 'smoke test'. Light a bundle of grass and green leaves with a match and throw it down the pit. After the flames die down smoke will be generated. This will pass up the pipe, and very little will be able to pass out of the open hole.

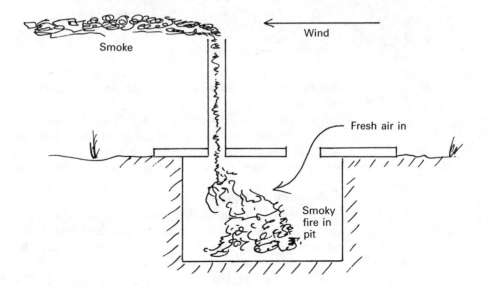

Comparison with ant turrets

The turret built by ants on the apex of an anthill is responsible for producing an almost continuous air movement through the underground chambers and passages of the mound, thus keeping them ventilated. The wind and the sun have the same effects on both the Blair vent pipe and the ant turret. It is possible to demonstrate this effect by closely observing a suitable ant mound with a prominent turret, and also locating one or more inlet holes at or near the base of the mound. As wind blows across the top of the turret air is drawn in through the inlet holes. Smoke applied just within the inlet hole will be drawn through the underground chambers and up the turret. A mosquito coil is ideal

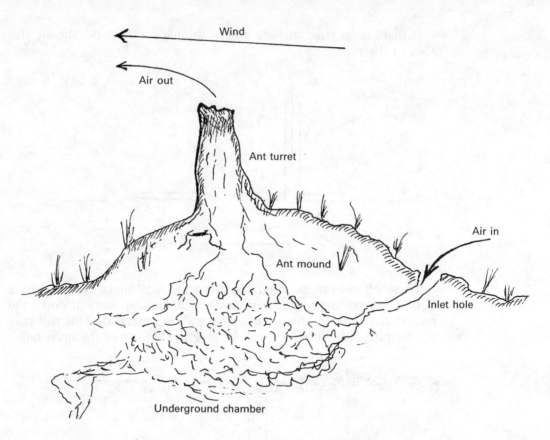

Ant turret acting as a ventilator for an ant mound.

for this purpose. This effect is maintained on still days when the turret wall is heated by the sun, causing air to rise in the pipe and thus ventilate the buried chambers. The Blair Latrine simply uses principles which have been employed successfully by ants for millions of years.

Efficiency of ventilation

The efficiency of ventilation varies with the type of pipe chosen and the direction of the structure in relation to the wind. If the latrine structure opening is facing into the wind, more air will pass through the latrine, compared with a latrine with an opening facing away from the prevailing wind. When the latrine opening faces the wind, air blows into the structure, and is forced up the pipe. This movement of air is assisted by the suction caused by wind blowing over the pipe. If the latrine opening faces away from the wind, the wind will try to draw air out of the structure, whilst the pipe is trying to draw air into the structure. When the pipe is efficient, the pipe will always draw more air up than the superstructure opening draws down, no matter what direction the structure faces. However if the pipe is less efficient, it is possible that in unfavourable conditions air may actually be drawn down the pipe to replace air passing out of the structure opening. This will lead to odours building up in the latrine, and is most common when brick vent pipes are built too small, and the wind is blowing away from the structure opening.

Latrine opening facing into wind.   Latrine opening facing away from wind.

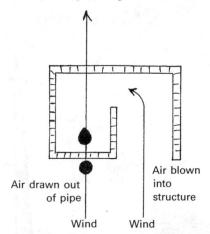
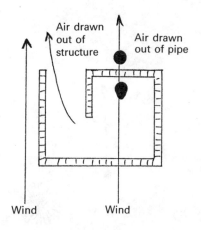

The most efficient pipes are made of steel, PVC and asbestos and are commercially available. Brick pipes are less efficient, but work well enough if they are made to required specifications. They should have an internal measurement of 225 mm × 225 mm (9 inch × 9 inch) and the internal wall should be smooth. The head of the pipe should be at least half a metre above the roof level, and preferably more. If a brick pipe is not tall enough, or is made with four bricks per column and not the recommended six bricks it will not work well. Also if the internal wall is roughly finished and the internal area is small it will not work well. Very often a builder leaves a lot of mortar on the inside surface, caring more about the external appearance of the pipe. In this case, the internal appearance is the most important and should be smooth.

It can be seen therefore that an efficient vent pipe is essential in the Blair Latrine. At one time, 150 mm diameter pipes were recommended as the best, but the high cost of these commercial units has led to the adoption of smaller pipes, the best available at the present time being made with 110 mm PVC pipe. Since most rural programmes use brick-built vent pipes, it is essential that care is taken in training builders to build the pipe correctly.

Another important aspect of the vent is that over the years cobwebs develop inside the pipe, and these can greatly reduce the efficiency of air movement. It is important to wash down the pipe with a bucket of water from time to time to keep the pipe clear.

Ventilation pipes do produce an odour which is quickly dissipated into the air. Obviously the higher the pipe, the less the smell will be noticed. It is wise to ensure that the air passing over the head of the pipe does not drift into a nearby house. In a typical homestead, the latrine is best placed downwind of the living area, so that the wind will carry odours away from the kraal, and fresh wind will blow into the superstructure opening. The illustration shows this clearly.

Although solar radiation is less important than other factors in deciding the orientation of the structure, it is wise to ensure that morning and afternoon sunlight do not shine too brightly on the internal walls of the latrine, as this may influence the emergence of flies from the pit. The latrine opening should face north or south if possible, or at least near

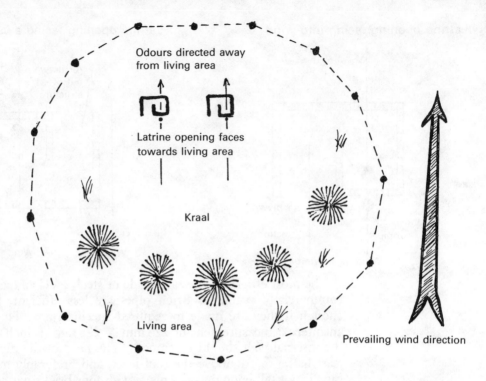

the north or the south. The priorities as far as latrine orientation are concerned should be as follows:

1. Privacy
2. Wind direction
3. North/South

The more efficient the pipe is, the less critical the wind direction and other factors become.

The control of flies

Blair Latrines are very effective at controlling flies. Indeed they were originally designed for this purpose. The theory of fly control is quite simple. Flies are attracted to a latrine by an odour and away from it by light. Once inside the pit, flies breed and when they emerge they fly towards the strongest light source, which in most pit latrines is the squat hole. Thus in most pit latrines odours come out of the squat hole, and flies pass freely to and fro through the same hole. There is no control of either odours or flies.

In ventilated pit latrines like the Blair Latrine, most of the odours are sucked up the pipe and escape into the atmosphere. Likewise most of the light falling into the pit passes down the pipe if the structure is fitted with a roof. Flies approaching the latrine from outside are therefore strongly attracted to the head of the pipe, and flies from within the pit are attracted up the pipe. These two groups of flies are never allowed to meet, however, because the head of the pipe is fitted with a corrosion resistant flyscreen across which they cannot pass. This makes it impossible for flies to enter the pit through the pipe, although some will find

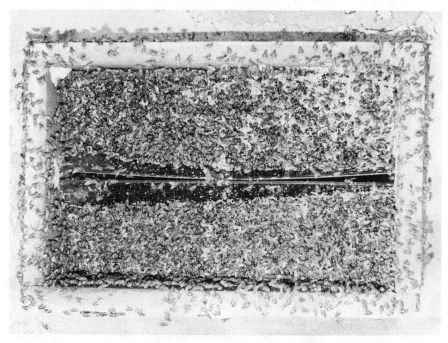

Early research of the Blair Latrine involved counting the fly output from ventilated and unventilated pit latrines. This trap is full of flies. As many as 140,000 could emerge from one pit in one year.

their way into the pit through the structure and squat hole. The number of flies attracted through this route will be small if the interior of the latrine is kept clean. However, if the interior is poorly maintained and smelly, flies will be attracted inside, and will find their way into the pit and breed there. This is one good reason why latrines should be kept clean. However, in the Blair Latrine, flies that escape from the pit will pass up the pipe and will not be able to pass the screen, and so become trapped. Such flies may pass up and down the pipe, but they are permanently trapped and die — falling back into the pit. They are the victims of their own instinctive behaviour.

If the screen is broken, all fly control is lost because flies will have direct access both into the pit (via the pipe) and away from the pit (via the pipe). It is clearly essential that the screen is kept intact.

Screen material

Most metal screens are destroyed rapidly at the head of the pipe because the gases passing up the pipe are very corrosive. In earlier times, most vent pipes in Zimbabwe were covered with PVC-coated fibreglass screens. However after four or five years this material became much weakened, and any strain on it could cause a tear. This type of screen is being phased out. Aluminium screens have also been tried with success, but the screen of choice is made of stainless steel. Aluminium and stainless steel screens are not manufactured in Zimbabwe, and many agencies donate screens to the Ministry of Health to assist in the sanitation programme. Currently there is a shift towards the use of aluminium as this is much cheaper than stainless steel.

Semi-dark conditions

The vent pipe can only act as a fly trap if the interior of the structure is kept semi-dark, thus allowing light to enter the pit predominately from the pipe. Flies move away from the pit to the strongest light stimulus, and this must be the pipe at all times. Originally, self-closing doors were fitted to Blair Latrines, but these often broke down, or when the self-closing mechanism failed, were left in the open position. This allowed light to enter the pit through the squat hole, with a resultant release of flies. Since fly control could not be guaranteed with such a system, the doorless spiral structure was designed, so that semi-darkness could be guaranteed, thus ensuring effective fly control at all times. Spiral shaped latrines can also be fitted with a door for increased privacy, but fly control is not lost if the door remains open. The spiral superstructure (either round or square type) is universally used in Zimbabwe, but elsewhere, where it is not used on VIP latrines, the fly control component is questionable.

The efficiency of the vent pipe as a fly trap was clearly shown in a simple series of experiments carried out during 1975. The short section reprinted beneath is extracted from a paper by the author written on the subject in the Central African Journal of Medicine, Vol. 23. No. 1. Jan. 1977.

1. *Proportion of flies passing towards the vent pipe.*
In this case a special trap was fitted within the vent pipe so that flies could ascend towards the flyscreen but collected in a side chamber if they died or came back down the pipe. A flytrap was also fitted over the aperture used for squatting. Flies were then released artificially from the base of the pit and the proportion of flies caught in the two traps was recorded.

Date	Location	No. released	Flies (pipe trap)	Flies (inner trap)
16–21.4.75	Chikurubi	197	171	0
22–30.4.75	Chikurubi	1,455	1,246	7 (0.5% of released)

2. *Proportion of flies emerging from used aperture.*
Flies were released artificially from the base of pits and the number trapped in flytraps fitted over the squatting aperture within the superstructure were counted.

Date	Location	No. released	No. trapped	% flies
16–21.4.75	Chikurubi	138	5	3.6
22–30.4.75	Chikurubi	1,406	66	4.6
2–6.5.75	Henderson	360	7	1.9
7–14.5.75	Henderson	170	4	2.3

3. *Natural fly emergence from vented and unvented structures.*
Four identical privies were built in a row, but only two were fitted with ventilation pipes, the other two, being unvented, behaved like normal pit latrines. These units were put into use for six months prior to the experiment. During the period October to December 1975, weekly counts of fly output were taken from one pair of structures at one time (vented and unvented) in order to enable the other pair to stabilise whilst in use. The traps were moved from one pair of privies to the other at monthly intervals.

Thus, during the period 8 October to 24 December, 13, 953 flies were trapped from the unvented structure, whilst only 146 were trapped from the vented structure. The structures differed only in the presence or absence of the ventilation pipe.

Period of trapping	No. trapped in unvented privy	No. trapped in vented privy
8 October to 5 November	1,723	5
5 November to 3 December	5,742	20
3 December to 24 December	6,488	121

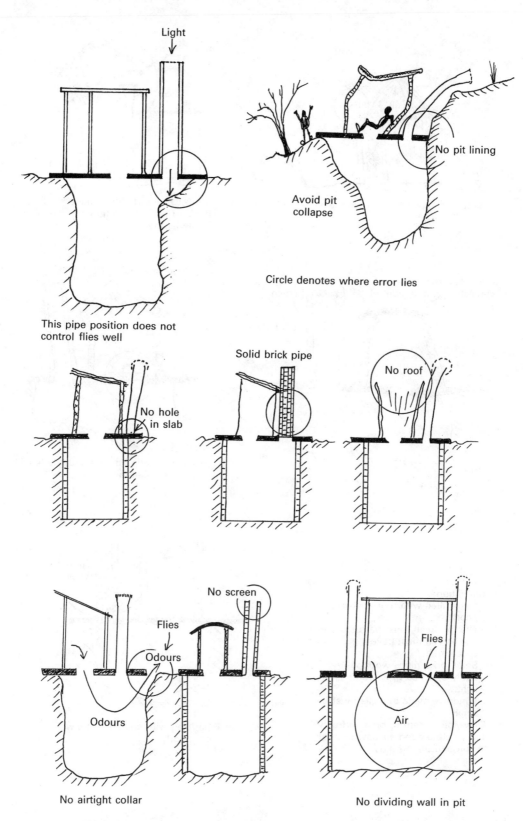

Blair Latrines that cannot work.

279

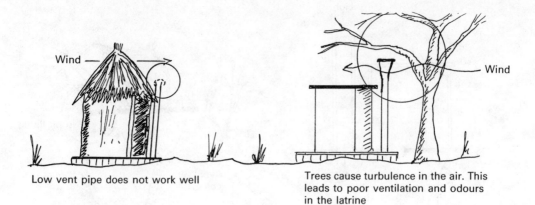

Low vent pipe does not work well

Trees cause turbulence in the air. This leads to poor ventilation and odours in the latrine

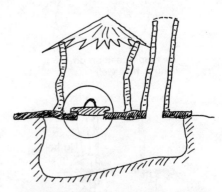

Bricks laid over pipe will stop ventilation and fly control

Closely fitting lid over squat hole will stop ventilation

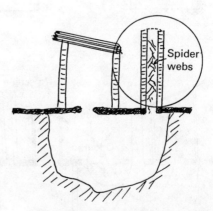

In addition:
  Pipes filled with mortar work poorly.
  The flyscreen must be rustproof.
  The screen must have a good open space area.
  Small brick pipes work poorly.
  Badly thatched grass roofs let in light.
  Windows made in walls must be small.
Allow the system to breathe freely and create a good air flow through the pit and up the pipe.

Pipe filled with spiders webs will work poorly

Blair Latrines that cannot work.

Since flies are attracted to light at the base of the vent pipe, it is important that light falling through the pipe can pass into the pit. Some builders construct pipes that are offset, and thus cannot direct light into the pit. These pipes do not control flies well. Since flies are attracted to odours, it is important that the only odours emanating from the latrine come from the vent pipe. There are several cases where this may not happen. If the slab is not sealed to the collar, for instance, odours will escape there and attract flies.

The efficiency of the vent pipe can also be reduced if it is placed near trees. It is essential that a free air flow is able to pass over the head of the pipe. If the laminar air flow is disturbed, and turbulence caused, the pipe cannot draw air well, and smells develop in the latrine.

Natural mechanisms

The remarkable similarity of ventilation in ant chambers and the Blair Latrine has already been described. This similarity is also seen in the behaviour of spiders when associated with either an ant turret or a vent pipe. There are frequent examples in both of where webs are woven near the apex of the vent in order to trap insects. Spiders spin webs within vent pipes for the same reason. Even more remarkable is the behaviour of lizards, which frequent the tops of vent pipes simply because they know that flies and other insects are attracted to that one place on the structure. Nature follows strict and dependable rules. Just

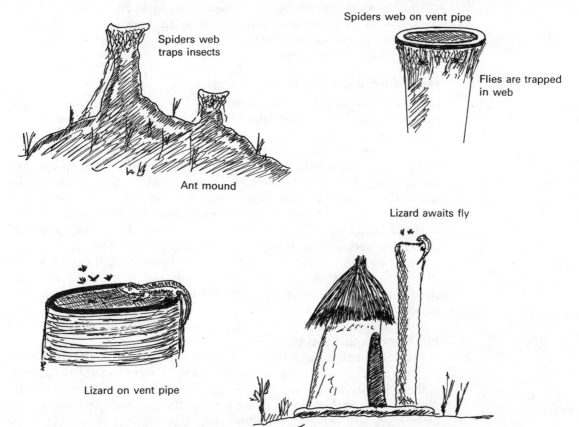

as the passage of air through the ant chamber and the Blair Latrine can be predicted with accuracy, so can the behaviour of flies, spiders and lizards. The dependability of the Blair Latrine can be ascribed to the fact that it uses mechanisms that have been operating in the natural world for millions of years.

Questions and answers

Do Blair Latrines control mosquitos?

Blair Latrines do not control mosquitos very well. Mosquitos are not influenced by light in the same way as flies. However, mosquitos only breed where there is a free water surface exposed in the pit. This happens normally in new pits which are dug where the water table is high, or older pits which are also in a high water table areas, but not used much for defecation. Some school latrines may enter this state at the end of the holiday period. Under normal conditions where the latrine pit is wet and used by a family, a 'crust' layer of decomposing material develops on the surface of the water or effluent layer, and mosquitos cannot breed in this. If the latrine is new or not much used for defecation, mosquito breeding can be stopped by adding about half a kilogram of polystyrene balls down the hole into the pit, so they float on the water. Mosquitos cannot pass through this layer of balls to contact the water, and breeding becomes impossible.

Small insects known as *psychodas* or midges can develop in wet latrines, and these are also common in septic tanks. They can play a useful role in the digestion of sewage. Fortunately they are neither a nuisance nor disease carrying.

How dark must the Blair Latrine be?

One complaint about the Blair Latrine is that it is too dark. Whilst it is true that semi-darkness is required for fly control, it is not essential that the latrine is as black as night inside. Provided the light coming down the vent pipe can act as a stimulus which attracts emerging flies, the interior can be illuminated by a diffuse light which comes in through the opening. It is also possible to place a small window in one of the side walls to allow a little diffuse light to enter, provided that it does not allow streams of light to pass directly into the structure.

If the latrine is built according to the correct dimensions it should not be too dark inside. However, some builders make the doorway too narrow, or paint all the walls and ceilings black, which can make the interior very dark. Also if the opening looks out over a dark area, like the shade of a big tree or building, the light entering that opening will be reduced, and the inside will be darker. There have been some cases where the inside walls have been painted white to lighten things up a bit. This is quite acceptable.

Problems may occur when the light of the sun shines directly on the inside wall of the structure in the afternoon. This reflected light may be sufficiently strong to attract flies out of the pit. However it should always be remembered that if the pipe is efficient, and the interior kept clean, very few flies will ever find their way into the pit.

Should a coverplate be used over the squat hole?

Sometimes a tin plate is laid over the squat hole, especially where the latrine is used as a wash room. This stops soap and other bathing utensils falling down the pit. It is seldom that a tin cover laid over the squat hole will fit tightly enough to affect the circulation of air greatly. If it does fit tightly, the air circulation will stop, and some of the properties of the latrine will be lost. It is not necessary or advisable to fit a tight fitting overplate on a Blair Latrine, which has been designed to operate without one.

How quickly will a family latrine fill up?

This will depend on many factors which include:

1. Size of pit
2. Size of family
3. How much water is added to pit
4. How much refuse is added to pit.

Under ideal conditions little refuse will be added to the pit, and paper or other degradeable material will be used for anal cleansing. The slab will be washed down daily with water, which keeps the pit contents moist. Moist faecal matter digests more rapidly that dry matter.

Under these ideal conditions, approximately $0.025\,m^3$ of waste will accumulate each year per user. This is slightly more than $1\,ft^3$ of solids.

Most Blair Latrines are built with a pit which is 3 metres deep although 4 metres is preferable. If we estimate that the pit is full when the contents are within $0.3\,m$ of the surface, the effective volume of the pit is $3.05\,m^3$ This is equivalent to about 120 man years of use, or twelve years use by a family of ten people, or twenty years use by a family of six people.

However, pits rarely operate under ideal conditions. Solid objects are often used for anal cleansing and water is not always added regularly to the pit. The latrine is also often used for garbage disposal. As a rule of thumb, if the pit is not well irrigated with water, and solid objects are used for anal cleansing, then the accumulation rate per person per year will be $0.04\,m^3$. A family of ten would thus get about seven years, and a family of six about twelve years life out of a $3\,m$ deep pit. It makes good sense to build pits deeper, say $4\,m$, and to use a well-made refuse pit for garbage disposal. Many Blair Latrines built in the mid 1970s for families are still in use (1989) and the pits are barely half full. A life of at least twenty years can be estimated for these old units. The frequent use of bathing water and the use of paper for anal cleansing are important factors which have contributed to the long life of these pits.

Should the latrine be used as a washroom?

Blair latrines make excellent washrooms, and they should certainly be used in this way. Any technique that improves family and personal hygiene should be actively encouraged. When used as a washroom, the latrine floor is washed down thoroughly, and this cuts down on potential odours caused by urine being absorbed into the cement floor. Water is also added to the pit, and this increases the rate of digestion of the contents of the pit. Wet pits last for longer than dry pits. Care should be taken however, to ensure that the amount of water does not overburden the pit. Some soils do not allow much moisture to be absorbed, and this must be taken into account on the individual site. The base of the pit can become clogged with waste material to the point where a well-lined pit can act as a tank, although this is unusual in most rural settings. If this happens, the pit will fill with water to a point which makes the structure unfit for use. The correct amount of water that any single pit can accept can only be judged by experiment by the users themselves.

If the latrine is to be used a great deal as a washroom, and ground conditions do not permit good water seepage, then it is advisable to build a 'tank' version of the Blair Latrine, which is described in another chapter. This technique also enables the owner to fit a flush toilet later if he wishes to 'upgrade' his latrine.

It is also most advisable to provide a convenient facility for handwashing within the latrine. Good handwashing practice is essential if the full benefits of an improved sanitation programme are to be gained. Preferably this should be some form of simple tap connected to a small reservoir. However this is often not very practical. The 'mukombe' may be used as an alternative technique as a hand-washing facility for rural areas. This is described in another chapter. Handwashing is a vital component of personal hygiene, and should be actively encouraged by all health workers.

How important is the latrine floor?
The latrine floor is the sloping component of mortar laid over the coverslab inside the structure. It is important that this is made with a strong mixture of cement and sand (ratio 3:1), and is carefully sloped towards the squat hole. This surface will be washed down many times in the life of the latrine and it is important that it does not erode away. If it is not sloped correctly pools of water or urine will be left standing, and these can make the interior insanitary.

Should a latrine floor be painted?
It is very common for urine to fall on to the latrine floor, and if this is not washed down regularly it will be absorbed deeply into the concrete work and will smell permanently. It is advisable therefore to wash down the floor daily and if possible to paint the floor with a material which is waterproof. A thick layer of bitumastic paint can help a lot, and this should be renewed every year. An even better material is black epoxy paint, which forms a hard surface, but this is expensive and not easy to buy. Frequent washing on a hard well-sloped floor is the best answer. Paint is unnecessary if this is practised.

How can faults be detected in the latrine?
Some newly built latrines smell and clearly are not working properly. The best way of beginning to find out what is wrong is to perform a 'smoke' test. This involves getting some grass, twigs and paper and wrapping them in a bundle. The bundle is lit with a match, and thrown

down the pit so that a great deal of smoke is made. If the latrine is performing properly, all the smoke should come out of the vent pipe, and none from the squat hole. If smoke comes from the squat hole, a mistake has been made in the construction. A common fault occurs in the vent pipe itself, which may be too small, and restricts the passage of air through the system. The latrine must be able to breathe freely, with air passages which allow a great deal of air to pass into the squat hole and up the pipe. The pit slab must be sealed over the pit collar, otherwise smoke may pass out of the space between them.

How tall should the vent pipe be?
Commercial pipes are usually about 2.5 m long. When builders build brick pipes there is often a great variation in the length. To work at its best the pipe should be at least a half metre above the roof level. Some districts build brick pipes which are one metre higher than the roof. These draw air very well. It is very unwise to save bricks by making the pipe shorter. The pipe must have an internal measurement of 225 mm × 225 mm (9 inch × 9 inch) to work effectively. The internal surfaces of the brick vent should be smooth, and the hole in the base slab made also 225 mm × 225 mm in size.

How many types of Blair Lactrine are there?
The Blair Latrine is characterised by the presence of an efficient screened ventilation pipe and a structure which guarantees semi-darkness within. Three basic types are built.

1. Single unit
2. Double unit
3. Multicompartment unit

The single unit can be made of grass, mud and poles bricks, ferrocement and other materials, with the brick technique being most common. Pipes can be made of reeds and mortar, bricks, PVC, steel and asbestos, with roofs being made of grass, concrete, tin and asbestos.

The double unit is normally made with bricks, with two units built side by side, often over a larger pit which is subdivided.

The multicompartment unit is also made with bricks, and normally built at schools. Single, double and multicompartment Blair Latrines can also be built over a tank and soakaway system, if the units are to be upgraded at a later date. These units are designed to accept more water, and can be desludged with conventional sewage tankers.

The first Blair Latrines were made with wood and had tin pipes. However, these were superseded in the mid-seventies with ferrocement structures equipped with asbestos vent pipes. Earlier doored models were replaced by the spiral model, and PVC pipes became more popular than asbestos pipes. Later low-cost models using traditional materials were tried, but these were phased out of the Ministry of Health programme because of their temporary nature. Brick structures equipped with brick ventilation pipes now dominate the programme. The 110 mm PVC pipe is the most common commercial vent pipe, and several models of the Blair Latrine are made commercially.

General maintenance

If the latrine is correctly sited on firm ground and built according to the recommendations, it should require very little maintenance apart from daily cleaning. The fly screen should be inspected from time to time, to ensure that it is undamaged, and the vent pipe itself washed down with a bucket of water periodically to ensure that it is free from spider webs. Damaged screens should be replaced, preferably with stainless steel screens which should last a very long time. It is important that the interior be washed down daily with water, as this keeps the inside fresh and the pit contents moist. Blair Latrines can make attractive feature in a garden, and many are surrounded by grass and flower beds. It is certainly worth painting and decorating the structure to make it attractive. Gone are the days when the family latrine was thought of as a menace at the bottom of the garden. A well-built Blair Latrine is a family asset.

Low cost Blair Latrines require more maintenance. This model is made from grass and has a pipe made of reeds covered with cement.

Upgrading the ordinary pit latrine

The common pit latrine normally consists of a hole dug in the ground, a covering slab made of wood, concrete overlaying wood, and some sort of structure built for privacy. Very often a door is used to ensure privacy. This simple structure is a very common sight in the Communal Lands and can work well if the hole is deep, the inside of the structure is dark and the slab is kept well washed down with water, and a cover plate is used to restrict access of flies to the pit. In most cases however, the pit is shallow, the structure allows a lot of light in and may not be kept clean and no cover plate is fitted. Even when a cover plate is provided, it is not always replaced immediately after use and flies still have access to the pit, and may leave the pit in large numbers once the plate is removed. Most pit latrines smell very badly and act as breeding places for flies. As many as 150,000 flies a year can breed and emerge from a single pit latrine. Such flies can carry disease and are a great nuisance in the homestead.

Fly breeding and odours can be overcome by the addition of an efficient screened ventilation pipe to the pit latrine. The development of improved pit latrines fitted with screened ventilation pipes and super-structures which guaranteed semi-darkness took place in Zimbabwe in the 1970s and these evolved into the Blair Latrine.

A great number of unventilated pit latrines exist and can be upgraded to make them odour and fly free by the simple addition of a vent pipe and a means of keeping the interior semi-dark. The structures are often well built with brick walls and are fitted with doors and a roof. Air holes are often built into the walls to aerate the interior. The pit itself is usually oblong and may or may not be lined with bricks. The great weight of a brick built latrine on an unlined pit can be disastrous unless the pit is dug in rock. Collapse is inevitable.

However in many cases bricks have been used to line the pit and a concrete cover slab has been fitted. If the structure is sound, such a pit latrine can be upgraded so that it works as well as a Blair Latrine.

Standard pit latrine.

The upgrading process

Pit latrines that can be upgraded should have pits which are brick-lined and only partly full. The cover slab should be made of robust concrete. The structure should be substantial and fitted with a door and a roof. The structure should be checked to see if the following steps are possible.

1. Fit a ventilation pipe with a flyscreen.
2. Ensure that the door can be adjusted so that it is self closing.
3. Replaster the floor so that it slopes towards the squat hole and is more hygienic.
4. Reduce the internal light within the structure by reducing the size of existing air holes in the wall.

The ventilation pipe

In the normal pit latrine air (and odours) from the pit must pass upwards and escape through the squatting hole. Flies are attracted to these odours and find their way directly into the pit and breed there. If a second hole is made in the latrine slab and a vent pipe is fitted through this so that the top of the pipe is well above the roof level, air from the pit will pass up the pipe and fresh air will be drawn down the squatting hole to replace it. The exact position of the pipe is not critical and it can be fitted inside the structure as well as outside. What is essential is that the pipe is fitted directly over the pit and that the pipe draws air efficiently.

In normal pit latrines which have not been designed for a pipe, this means that a hole for the pipe will have to be knocked through the concrete base slab within the structure. This can be made either behind the squat hole or to one side of it. When pipes are fitted in this way, a matching hole should also be made in the roof, through which the pipe can be fitted. The minimum size for an efficient vent pipe is 110 mm, the most convenient being PVC. The screen should be made of aluminium or stainless steel. PVC and asbestos pipes are available commercially. The hole through the concrete slab should be cut out to the required size with a hammer and cold chisel. It is wise to ensure that the slab is strong enough to accept this treatment. The hole should pass right through the slab. The vent pipe is attached so that it fits neatly and tightly into the hole in the concrete slab. Some cement mortar can be trowelled around the pipe where it fits into the slab and also where it passes through the roof.

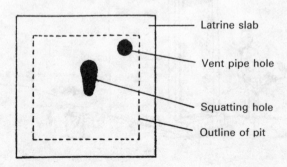

The door
Many normal latrine doors can be left open. If this is so, flies emerging from the pit will not choose to pass up the pipe but will pass through the squat hole. Under these conditions there is no control of flies. It is important, therefore, to design the door so that it is self-closing. When the interior is darkened, when the door is closed, flies will be attracted to light passing down the vent pipe and will be trapped there. One technique which works well is to cut out a piece of rubber tyre and attach it to the frame of the door in such a way that it bends backwards when the door is opened. The rubber can be nailed or screwed into the door frame. It is necessary to experiment with this technique to ensure that the door will self-close. In fact there are several techniques using rubber that might be useful. Spring hinges can also be used, but are not easy to obtain and require regular maintenance.

The latrine floor
Many latrines smell because the floor itself is soiled or the cement is filled with stale urine. All latrines should be built with a hygienically shaped floor with all surfaces steel floated and sloping towards the squat hole. The mortar for the floor should be made with cement and clean river sand using a mixture of 1:3. Foot rests can be fitted but are not essential. Ideally the cured cement floor should be painted with a thick layer of waterproof paint like bitumastic, but this is not always available. The most important aspect of the floor is that it is frequently washed down with water. This helps to keep the interior of the latrine fresh and the pit contents moist.

Semi-darkness
In order to make fly control effective, the interior should be semi-dark. This does not mean that the interior should be as dark as night however. Flies will be attracted from the pit to the greatest light source, which will normally be the vent pipe when the superstructure is semi-dark within. When upgrading existing latrines it may be necessary to paint the inside of the roof black. When seen from the depths of the pit, a fly is normally attracted to light coming from the inside of the roof. Air vent holes made in the walls may require reducing in size, but can remain in use. Fly control depends on a combination of factors. A good vent pipe, a clean interior and semi-dark conditions are essential.

Routine maintenance
In upgraded latrines it is important to ensure that the self-closing door continues to work and that the floor is washed down frequently and the vent pipe is fitted with a complete fly screen. Fly control is lost if the screen becomes broken as the flies will pass across the broken screen, down the pipe, into the pit and up the pipe away from the pit. Over the years it is possible that cobwebs will grow inside the pipe. If webs become too thick they can seriously reduce the rate of air flow in the pipe, and this will lead to odour and fly problems. It is therefore wise to flush a bucket of water through the pipe from time to time, to clear the webs.

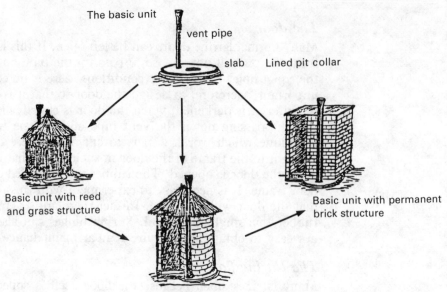

Upgraded pit latrine.

If the latrine is used as a washroom, the floor will slowly erode away over the years and may require replastering. If these simple maintenance procedures are followed, the latrine will provide good service for many years.

Essential features of the Blair Latrine:

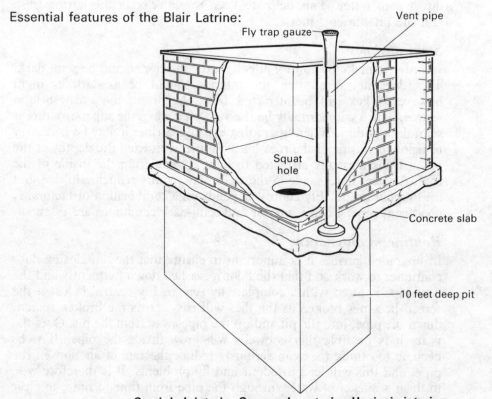

Semi dark interior Screened vent pipe Hygienic interior.

How to build a Blair Latrine

The single compartment version

The single compartment brick-built Blair Latrine is the most popular and the most common version built in Zimbabwe. Most are built with brick pipes, and have ferrocement roofs. It is also possible to fit a PVC pipe, and the roof can be made with any permanent material such as corrugated iron or asbestos. Although many latrines of this type are built with donor assistance, it is possible for a family to embark on a programme of construction stage by stage by itself without assistance. If funds are limited, it is best to put money into the concrete cover slab, the brick collar and a permanent pit lining first. A temporary superstructure made with sun dried bricks, reeds or grass can be built over this, as a first step. At a later stage when the family is able to raise more funds, a more permanent structure can be made. However if funds are available it is best to go right ahead and make a permanent structure right from the start which should last a small family for a generation.

The most common type of Blair Latrine in Zimbabwe — a single compartment brick structure with a brick vent pipe.

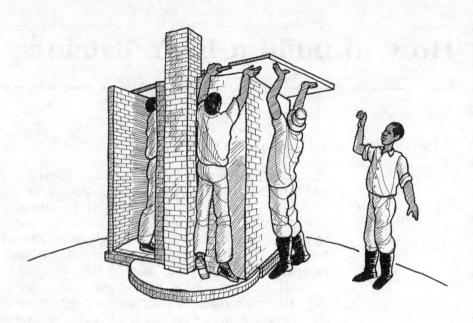

The construction

Tools required
Pick, shovel, bucket, wooden and steel floats, tape measure, rope, trowel, spirit level, string and hammer.

Building materials required
1. Cement 5–6 pockets
2. Bricks, approx. 1000–1200
3. River sand (approx. ½ m³)
4. Pit sand (approx. 1½ m³)
5. Gravel (1/8th m³)
6. Reinforcing wire (25 m × 3 mm)
7. Chicken wire (1.7 m × 2.0 m)
8. Flyscreen

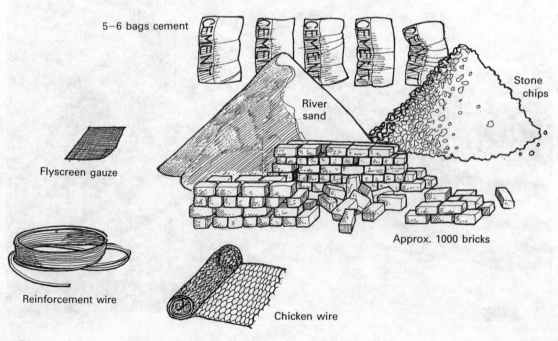

Siting

The best place for a latrine is downhill and at least 30 m away from a well. It should be sited somewhere near the house, and preferably downwind of it, with the opening facing the house. The soil should be firm and on a slightly raised place, so that rainwater will drain away. It should also be in a place where there are no trees to disturb the airflow across the top of the pipe.

The pit

Once the latrine has been correctly sited, a 1.5 m diameter circle is made on the ground to mark the shape of the circular pit. A circular hole is now dug in the ground 5 ft (1.5 m) in diameter and at least 3 m and preferably 4 m deep. The sides of the pit should be straight, and lined with cement mortared bricks from the bottom to the top. The mixture for mortaring is 8 parts pit sand to 1 part cement. The base of the pit is left unlined.

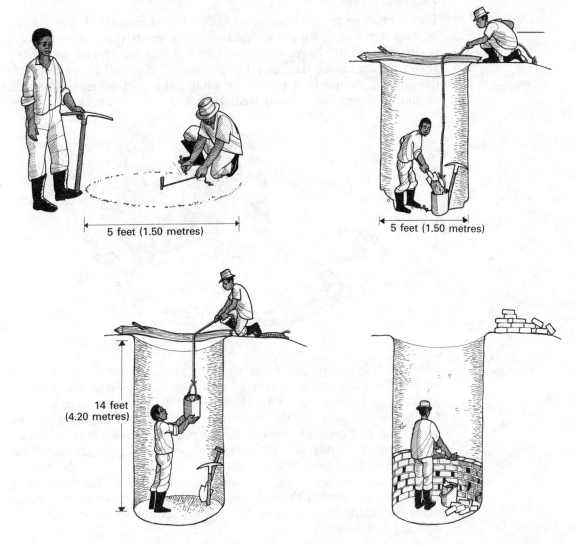

The collar

A ring of bricks, 75 mm (3 inch) deep and 225 mm (9 inch) wide should be laid in cement mortar around the head of the pit, and levelled off to accept the cover slab.

The coverslab

This is made to the measurements shown in the diagram. It is 1500 mm in diameter and 75 mm thick and requires about three quarters of a pocket of cement. If gravel is available the mixture should be 3 parts gravel, 2 parts river sand and 1 part cement. If gravel is not available the mixture should be 5 parts river sand and 1 part cement. The sand should be sharp and washed well. It should not contain dust.

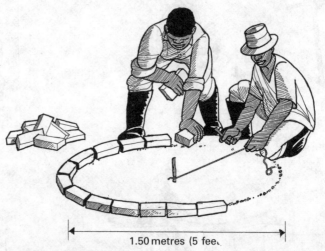

1.50 metres (5 feet)

The slab is constructed by arranging a 1.5 m diameter ring of bricks on levelled ground which is used as a mould. It is best if a piece of plastic is laid down first. Two holes will be made in the coverslab, one for the pipe and one for the squatting hole. These are made by placing suitably shaped templates within the slab mould as shown in the diagram. The size of the vent hole will depend on the type of pipe chosen. If this is a 110 mm PVC pipe, the pipe itself can be used to make the hole. If it is a brick pipe, two bricks can be laid down to leave a hole measuring 225 mm × 225 mm in the slab. Suitably shaped bricks can be used to make the shape for the squat hole which should about 150 mm wide and 300 mm long.

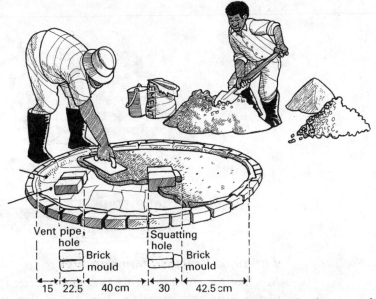

Once the mould is set up, some 3 mm reinforcing wire is cut and laid out to form a grid with wires 100 mm apart. Once the wires have been cut they are laid on one side and the concrete mixture made up. Half the mixture is laid in the mould — the reinforcing wires are then placed in position, and the remaining concrete added and trowelled flat. The slab should be covered with grass or sand and allowed to cure for at least five days. It should be kept wet throughout this period.

Fitting the slab

After curing, the slab should be carefully lifted, washing and laid over the collar of the pit in a bed of cement mortar. The slab should be arranged so that the future opening faces the right direction for later use. The latrine will work best when the superstructure opening faces into the prevailing wind. However, it is also important to consider privacy. Normally the latrine will be built downwind from the dwelling with the structure opening (and also the vent pipe) facing towards the dwelling. The slab is rolled into position over the collar and embedded in cement mortar which binds it to the collar and makes an airtight fit. It is important that the slab and collar are fitted together neatly.

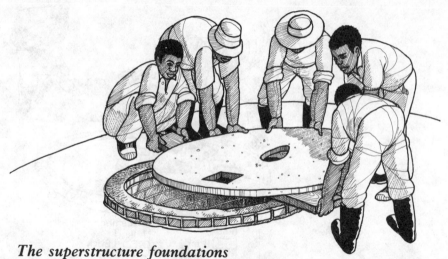

The superstructure foundations

Much of the superstructure is actually offset from the pit, and only part of it lies on the coverslab. The rest of the structure lies to the side of the pit and rests on a prepared brick foundation. At this stage it is necessary to decide on the shape of the structure — a round spiral or a square spiral. Neither is fitted with a door, although a door can be fitted as an optional extra. The foundation is made up of a line of bricks laid flat so that their width is 225 mm (9 inch). The foundation is built up so that it lies level with the coverslab.

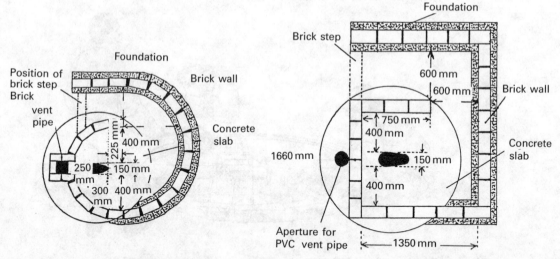

Plans showing round and square spiral superstructure with foundations.

The superstructure

A 112 mm (4½ inch) wide brick wall is built up on the foundations and the coverslab to the dimensions shown in the diagrams. The square structure is becoming more popular since it is more spacious inside than the circular structure and more suitable for bathing. The round structure uses about 400 bricks, the square structure, 600 bricks. Cement mortar is used for the construction. It is best that the bricks are fire-cured for a long life. Sun-dried bricks are less permanent even when covered with mortar. The internal walls must be plastered to make the latrine more hygienic, and easier to wash down.

If a brick vent pipe is built, this forms part of the superstructure. Each course must be built with 6 bricks to give an internal dimension of 225 mm × 225 mm as shown in the diagrams. Internal surfaces of the pipe wall should be smooth. The pipe should be built at least half a metre above roof level. The vent pipe is discussed later.

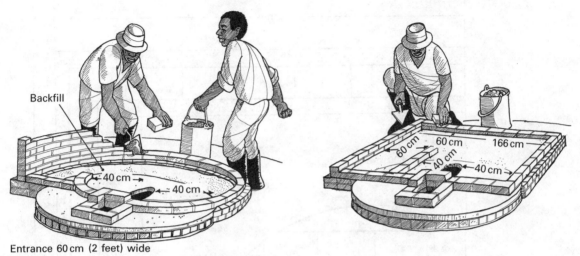

Round spiral.   Square spiral.

The latrine floor

A carefully shaped latrine floor is essential for hygienic purposes. Before the floor is laid a brick step is made at the entrance to the structure as indicated in the diagram. One brick height is sufficient. The space between the concrete slab and the brick foundation should be filled with chips, gravel, half bricks or well-packed soil up to slab level. The concrete latrine floor is laid within the structure. Since this area will be used constantly and considerable amounts of water will be washed over it as the years pass, it should be made of high strength concrete. A mixture of three parts river sand and one part cement is ideal. The mortar is laid down from the entrance step to the squat hole along a slope, to aid drainage during washing down. The floor must have a hard surface and the final finish can be steel floated. Footrests can be added into the floor slab, but these are not essential. If the floor is not sloped and dished, pools of urine will build up inside the latrine and become odourous.

The latrine floor is an important part of the structure and is a hard working surface which should be washed down daily with water.

The floor is sloped from the brick step at the entrance down towards the squat hole. It is important that washing water drains completely into the central squat hole.

It is possible to build a pedestal arrangement suitable for sitting over an enlarged squat hole. This can be made of bricks and cement and fitted with a wooden cover if necessary.

Note: On most Blair Latrines the concrete base slab is made 1500 mm in diameter. This is placed in position over a pit 1200 mm wide and at least 3 m deep. The preferred depth is 4 m where a longer life is required from the pit. The diameter of the slab can be increased to 1800 mm and the diameter of the pit to 1500 mm. This will require more concrete, but the pit will last at least half as long again.

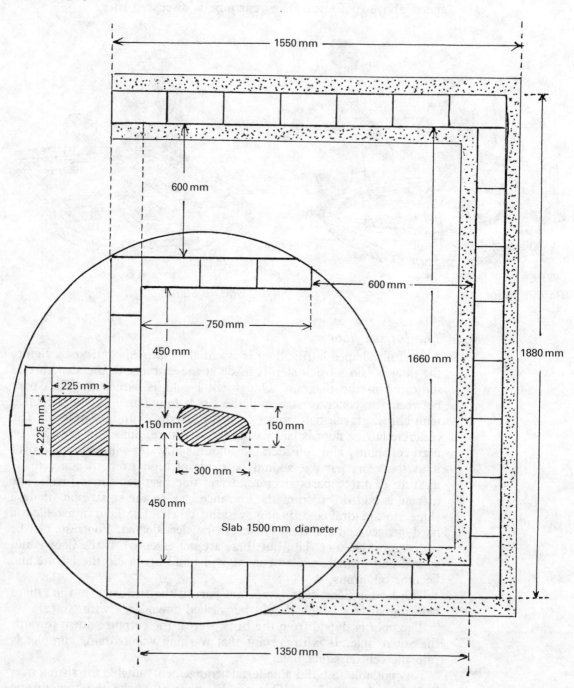

Dimensions of slab and structure (mm).

The latrine floor.

The latrine floor is a hard-working surface and should be made with a strong mix of concrete.

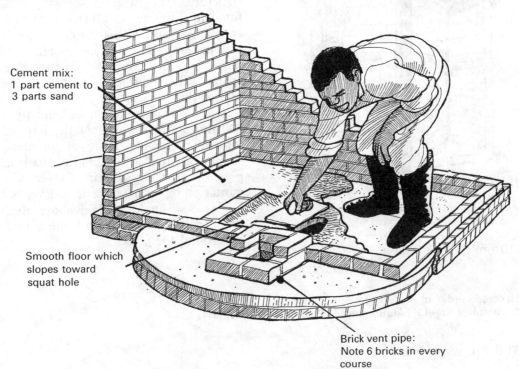

Cement mix:
1 part cement to 3 parts sand

Smooth floor which slopes toward squat hole

Brick vent pipe: Note 6 bricks in every course

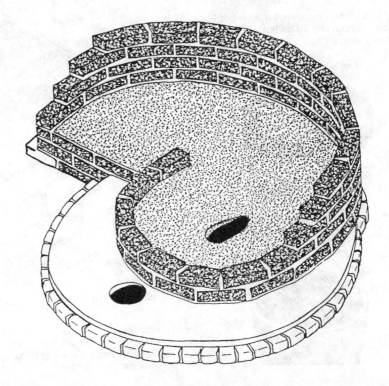

The ventilation pipe

This is a vital part of the latrine and if made of bricks is incorporated into the wall of the structure. The diagram shows a PVC pipe fitted to a latrine. In a brick vent pipe, each course is built with 6 bricks. The internal wall of the pipe should be smooth and have a dimension of 225 mm × 225 mm. Rough internal walls interfere with the airflow, and reduce the efficiency of the pipe.

The brick pipe should be built up at least 500 mm above the roof of the latrine. A fly screen, preferably made of stainless steel, should be fitted to the head of the brick vent pipe by embedding it in a layer of cement mortar on top of the uppermost layer of bricks. No bricks should be laid above the screen layer, as these interfere with the movement of air around the screen.

110 mm PVC vent pipe.

Flyscreen order of preference:
1. Stainless steel 2. Aluminium 3. PVC coated fibreglass.

The PVC vent pipe.

Note: This should be 2.5 m long and at least 110 mm diameter. Screen should be firmly attached. (Screen size: 180 mm diameter).

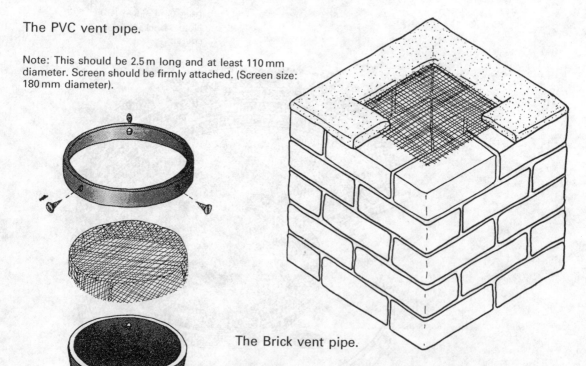

The Brick vent pipe.

Note: Six bricks are laid per course. Internal surface of vent is smooth. Vent is built half a metre above roof level. Screen is fitted on flat top of vent. (Screen size: 300 mm × 300 mm).

Brick vent pipe with arched flyscreen.

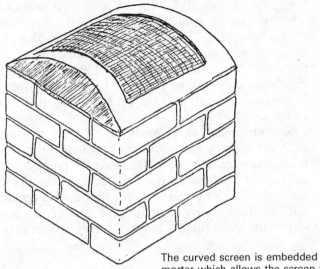

The curved screen is embedded in a layer of cement mortar which allows the screen to be observed easily.

150 mm diameter asbestos vent pipe fitted to a round spiral superstructure.

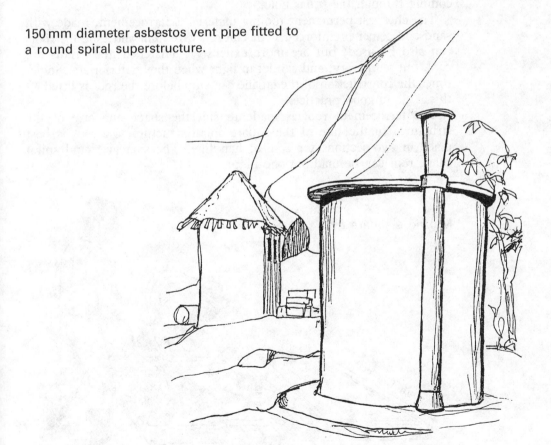

This is the most effective and durable ventpipe being little effected by sun, rain or gases.

Drawing by Hannah Schreckenbach.

Several other pipes can be fitted to Blair latrines, the most common being the 110 mm PVC vent made by Prodorite, Ltd., Harare. A good asbestos pipe is available from Turnall. Metal pipes tend to corrode after a few years. Commercial pipes are fitted with flyscreens, the best being of stainless steel. Larger pipes draw more air, but cost more — the smallest recommended pipe is 110 mm in diameter. The ventilation pipe must always stand proud of the roof. Obviously the vent pipe hole in the concrete slab over the pit should lie directly over the pit and not to one side.

Smooth-walled pipes made of PVC or asbestos are more efficient than brick pipes and more consistent in their dimensions. Obviously brick pipes are more desirable from the point of view that the materials are readily available in the rural setting. On the other hand it is very easy for a builder to make a vent pipe that looks good from the outside, but functions poorly. Where brick pipes have been chosen, it is very important that builders know the correct methods of construction.

The roof

All Blair Latrines must be fitted with a roof. Without a roof, the vent pipe cannot act as a flytrap, since flies will be attracted to the light coming through the squat hole.

The cheapest permanent roofing material is ferrocement, made with sand and cement reinforced with chicken wire. Tin and asbestos roofs can also be used, but are more expensive. Thatched grass roofs are cool, but temporary, and can let in light when they deteriorate. Sometimes the construction of the latrine can stop before the roof is fitted — this is not a good practice.

The ferrocement roof is made to suit the shape and size of the structure. In the case of the square spiral structure, the roof is best made in two sections for ease of handling. The smaller round spiral shape roof can be made in one piece.

Making a square roof.

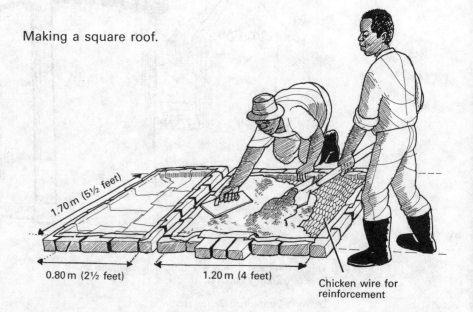

Construction method for roof

The roof is made by levelling off a piece of ground on which the roof is to be cast. If possible lay down a piece of plastic. The size of the structure is measured and the roof is made so that it overlaps the brick structure by 50 mm all round. The shape is marked on the ground. The diagrams indicate the approximate shape, although this will vary slightly from one structure to another.

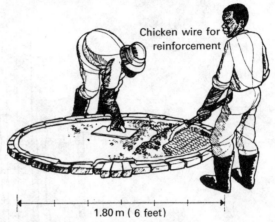

Making a round roof.

A line of bricks is laid on the ground in the shape of the roof. In the case of the square roof, two separate brick moulds can be laid down. If plastic sheet is not available, a thin layer of sand should be laid down within the bricks, to make the roof easier to lift later.

A piece of chicken wire is now cut to the shape of the roof. 25 mm (1 inch) chicken wire is best but 40 mm (1½ inch) can be used with care. This should be trimmed to fit inside the bricks.

The concrete is now mixed in the proportion 1 part cement to 3 parts sharp well-washed river sand. A firm slurry should be made. This will take about one bag of cement for the larger roof, and a little less for the round roof. Make enough mix for the complete job. Add half the mixture into the brick mould and level off with a wooden float. This first layer should be about 12 mm (½ inch) thick. Add the chicken wire and level this. Now add the rest of the mixture over the chicken wire and smooth level. The total thickness of the roof should be about 25 mm – 30 mm.

Once cast, the roof should be covered with grass or sand and allowed to cure for a period of one week before it is moved. It should be kept wet all through this period. After the curing period, the bricks should be removed and the roof rotated on the ground before it is lifted. Great care should be taken in lifting the roof. Then it should be fitted on to the structure and mortared in position.

Finishing off

Soil should be built up around the latrine so that the ground is raised slightly above the surroundings. This helps prevent erosion around the latrine during the rains. Planting grass helps to consolidate the soil around the latrine. Latrines are more durable if they are plastered with cement mortar. Plastering the inner side of the structure is essential for hygienic reasons. If sun-dried bricks are used, plastering is essential inside and outside. Fired bricks are preferable. Latrines can be painted to improve their appearance, and a neat pathway laid to the doorway.

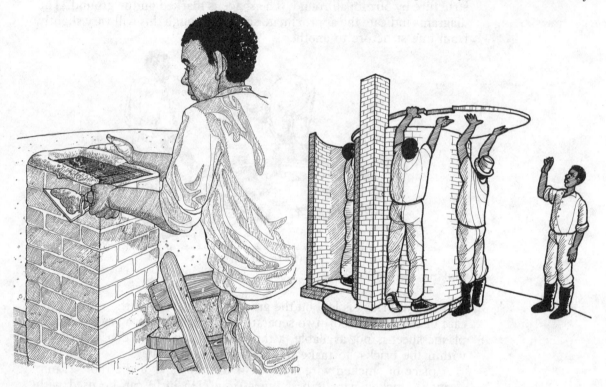

Basic requirements for Blair Latrines

1. It is important that the pit is deep — at least 3 metres and preferably 4 metres. The walls should be cut vertically and supported with a brickwork lining.
2. A brick collar is important and should make an airtight seal with the cover slab.
3. The structure opening should preferably face into the prevailing wind. It helps if morning and afternoon sun do not reflect from the walls into the interior of the structure.
4. A roof should always be fitted.
5. The brick vent pipe must be made to the specified size, and fitted with a corrosion resistant flyscreen.
6. Vent pipes should be built or fitted directly over the pit and not to the side, so that light can penetrate down the pipe directly into the pit.
7. That the latrine is built on an elevated site, not in a hollow.
8. The floor should be sloped and made of very strong concrete, and washed down regularly.
9. The screen should be inspected periodically.
10. Internal walls of the latrine should be plastered.

The brick-built Blair Latrine with a brick vent pipe is the most common Latrine in the communal lands.

Drawings by Hannah Schreckenbach.

The Blair Latrine in its natural habitat.

Matabeleland version

For several years the Office of the PMD in Bulawayo has promoted a version of the Blair Latrine known as a 'Brick Flu'. This is very similar to the standard Blair Latrine built with a brick pipe, but the pipe is formed under the roof level by two walls of the structure, closed off with a slab of concrete. This makes a triangular pipe. Above roof level the pipe is square and elevated at least 0.5 m above the roof.

This model is rugged and effective and many thousands of units have been made with the square spiral configuration.

A great deal of emphasis is placed on the use of concrete slabs, not only to make the pit slab and roof, but also to close off the chimney. The design overcomes the problem encountered in many brick built pipes, namely roughness of the internal walls. It is necessary however to build slabs of an exact size so that the pipe does not leak and draws air well.

In some areas of Zimbabwe, builders appear to design their own ventilated pit latrines, often disastrously. There is a need therefore to standardise the design, and very specific design details have been circulated by the Blair Laboratory for many years. The 'Brick-Flu' design is a case where the Public Health Inspectorate have been strict about design and the technique used in construction. This approach has clearly worked, since nearly all 'Brick–Flu' latrines function properly and have few defects.

The square interior can be used as a washroom, but very often a shower room is built on to the latrine structure, so as to form a double

'Brick-flu' improved ventilated pit latrine.

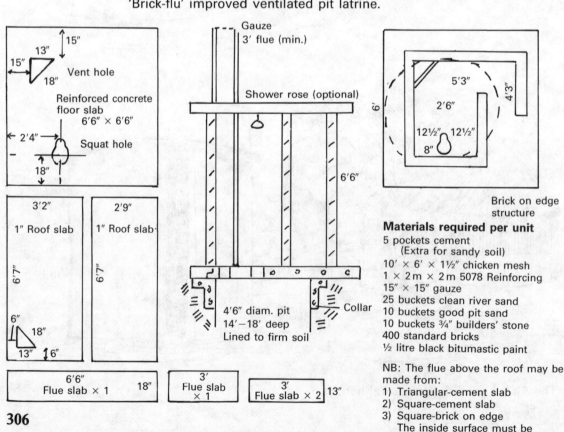

compartment structure. Waste water from the washroom drains into the pit under the latrine. The vent pipe for the 'Flu' can be very high (up to 3 metres) and therefore draws air very well. It is now established that short stubby pipes do not draw air very well — tall ones with smooth internal walls work much better.

Structures can be built in standard brickwork with concrete blocks or even in stone. To economise with bricks, they can be laid on edge. In sandy areas where bricks cannot be made directly from the soil, sand can be mixed with cement (12 parts sand to 1 part cement) and moulded to form bricks which can be used to line the pit and build the structure. In firm soils the brick lining may not need to go the base of the pit, and can be recessed in the upper strata. In loose soils the brick lining must go to the base of the bit. Pits vary in depth, but the deeper they are the better, as the life of the unit is extended.

Details of the 'Brick-Flu' are given below as produced by the PMD, Matabeleland.

Upgradeable version

Where a family wishes to begin constructing a Blair Latrine and does not have the money to complete a permanent structure it is possible to start a structure by making the pit lining and slab in bricks and concrete and adding a grass or mud structure to this which can be upgraded and made more permanent later. The best vent for this type of upgradeable construction is a PVC pipe with a minimum diameter of 110 mm. This should be fitted with a corrosion resistant flyscreen, preferably made of aluminium or stainless steel.

The technique used for the pit and the coverslab is the same as for the single brick Blair Latrine. The slab should be made in concrete, with a ventilation hole suitable for the vent pipe chosen. The coverslab is cemented on to the pit collar and the vent pipe fitted, or built in brick if this is the type chosen.

This *substructure* forms the basis of all further constructions, whether they be temporary or permanent.

The superstructure

This can be made in grass, reeds, mud brick, or if permanence is required, fired brick.

Reed and grass structure

This is temporary, but if well built can last for several years, and is cool inside. It can always be upgraded at a later date. A series of stout reeds or poles are cut approximately 2 metres long and embedded in the ground around the collar. These can be laid so as to make a spiral doorless structure or a non-spiral structure with a door. Once the skeleton of the structure has been made, these are strengthened by fastening a series of split reeds horizontally around the structure. This framework is filled with reeds or grass to make a thicker wall. A grass roof is fitted over the structure, so that direct light cannot penetrate from the sky. The vent pipe should pass through the grass roof so that light cannot pass through the hole into the structure.

The base slab should now be extended to make a good latrine floor which extends to the walls of the structure. The slab should be made so that it is dish-shaped and when it is washed down with water, drains into the squat hole.

This structure will ventilate well, although most of the vent pipe lies within the structure. Since brick pipes are large, tubular pipes made of PVC, asbestos or steel are more suitable. The vent pipe should extend well above the level of the roof.

If a doored structure is chosen, this must be sprung with a rubber tyre hinge so that it will always close automatically. Spiral structures are best for providing guaranteed conditions of semi-darkness within the structure.

Upgrading the superstructure

The same slab and pipe can be upgraded at a later date to form a more permanent structure. The reeds and thatch can be removed and replaced by a fired brick structure and a permanent roof. The instructions for the single brick Blair Latrine can be followed. The actual size and shape of the structure is not important so long as it is semi-dark and is easily washed down and maintained. The illustrations below show how the upgrading process works.

Upgradeable Blair Latrines.

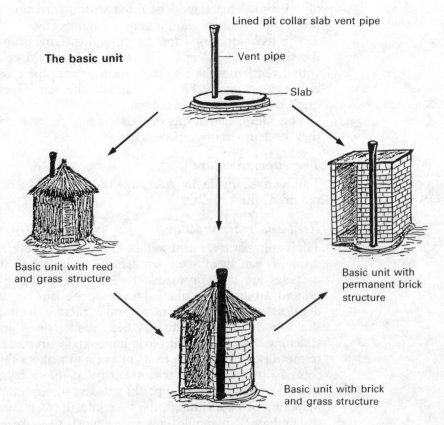

Ferrocement spiral version

This spiral ferrocement model of the Blair Latrine was first introduced in 1976 and replaced the earlier ferrocement doored version. It is ideally suited to areas where relatively large numbers are required to conform to a standard pattern. The commercial form is an example. The concrete base slab, superstructure and ventilation pipe should all be standard units, and the properties of the structure can be guaranteed. The structure is relatively light compared with bricks, and is not offset from the pit, but is made directly on the base slab, which is 1900 mm in diameter.

One of the most successful models of Blair Latrine — the ferrocement spiral fitted with a large asbestos vent pipe.

Brick and ferrocement spiral Blair Latrines fitted with 150 mm asbestos and PVC vent pipes.

The pit

This is 3 m to 4 m deep and dug 1.75 m in diameter so that when it is lined with bricks the internal diameter is 1.5 m. A collar made of a ring of bricks is mortared in position at the rim of the pit. The base of the pit is not lined, but left for seepage.

The base slab

This is cast 75 mm thick and 1900 mm in diameter with a mixture of 4 parts 12 mm stone, 2 parts river sand and 1 part cement. It is reinforced with 3 mm wire placed in a grid pattern with 100 mm spaces. The squat hole and vent hole are made in the slab by inserting templates within the slab mould. The vent hole must be suitable for the pipe chosen which will normally be made of asbestos or PVC. The measurements are provided in the illustration overleaf.

Base slab.

Showing mould and hole positions.

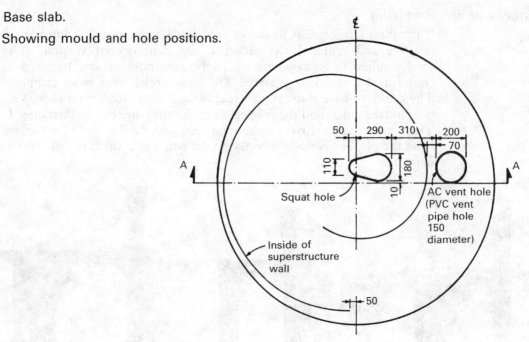

(110 mm PVC pipe can replace 150 mm asbestos pipe.)

The superstructure

The slab is moved on to the collar so that the opening of the final structure will face either north or south, or into the prevailing wind. A suitable mould for building the ferrocement structure is sold by McDiarmid & Co, Harare and is made with corrugated steel sheets. The 'form' is made in four main parts which are fitted together with wedge fasteners. These are erected over the slab and positioned as shown in the diagrams, so that the vent pipe lies outside the structure and the squat hole lies within the structure.

A layer of 38 mm (1½ inch) chicken wire (width 6 feet) is stretched around the mould and secured top and bottom and half way up the structure with 3 mm wire. The wire is made to fit tightly.

A mortar mix is now made up. This is 5 parts sand and 1 part cement. The best sand is a mixture of 50% river sand 50% pit sand. Alternatively a good 'builders' sand' with a mixture of coarse and fine particles can be used. The mixture should have enough fine material in it to make it stick to the mould, and enough coarse material to make it strong when cured.

The cement mixture is applied with a wooden float and forced into the mesh all over the mould, so as just to cover the mesh layer. This first layer of mortar is left to set for two hours and a second layer is added and built up to give a total thickness of 40 to 50 mm. After two days in warm weather, the formwork is disconnected and taken apart, and removed to leave the ferrocement structure standing. This is kept wet for one week to give strength. The inside of the structure is plastered to give extra strength and a smooth internal wall. A line of bricks make a step at the entrance and a latrine floor is laid within the

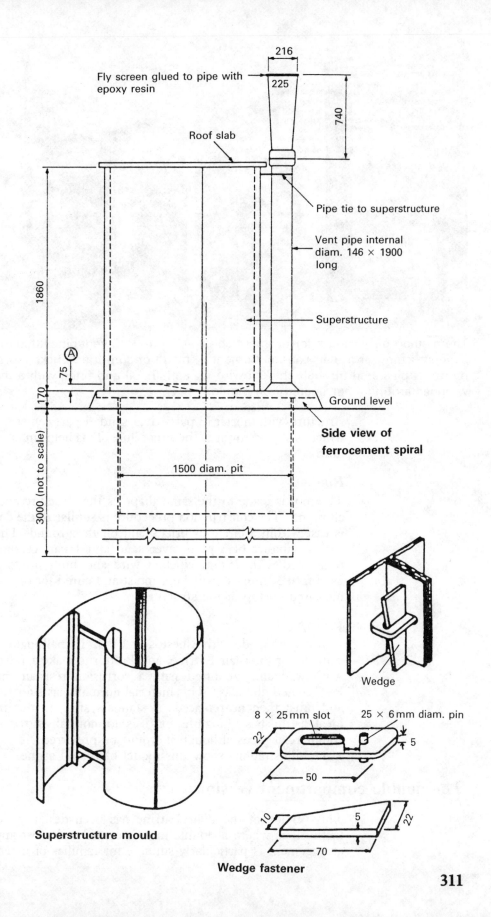

The method of making a ferrocement spiral superstructure for a Blair Latrine. Cement mortar is plastered through chicken wire wrapped around a corrugated iron mould.

110 mm PVC vent pipe fitted to a Blair Latrine made of sun-dried bricks mortared into anthill soil and fitted with a thatched roof.

structure with mortar (3 parts river sand: 1 part cement) to form a floor which is sloped towards the squat hole. This helps in cleaning down the unit.

Roof

The roof is made to the same shape as the structure and 50 mm larger all round. The construction procedure is similar to the coverslab, which is cast within a circle of bricks laid on the ground. The slab is made with a mixture of 3 parts river sand and 1 part cement. This mix is reinforced with 25 mm chicken wire and built up to a thickness of 25 mm to 30 mm. This is kept moist and cured for one week, and then mortared on top of the structure.

Vent pipe

This can be made with asbestos or PVC. Turnall make a 150 mm pipe suitable for the Blair Latrine, and Prodorite make a 110 mm pipe. Both work well and are fitted with a corrosion resistant mesh, normally PVC coated fibreglass. This material normally lasts for four to five years and must then be replaced. A stainless steel or aluminium screen is ideal, as this has a long life, but it is not normally fitted as a standard to commercial pipes although it would be preferred.

The illustrations show this model of Blair Latrine.

The double compartment version

This version of the Blair Latrine has been designed for families and institutions where a double unit is considered most appropriate. The arrangement is particularly suitable for families or extended families.

The square spiral brick Blair Latrine with a 110 mm vent pipe. This is a tank version — suitable for peri-urban settlements.

A double compartment Blair Latrine made with bricks.

It is important to remember that each Blair Latrine unit operates as a single unit from a functional point of view. If the latrine is to remain odour-free one vent pipe must serve only one pit and one squat hole. In the case of the double or multicompartment version it is essential that the pit is subdivided so that air cannot pass from one pit to the next. If there are holes in the subdividing walls of a double or multi-compartment pit, air will be able to pass down one squat hole, through the dividing wall and up the next squat hole. The pipe may have little influence on the air flow. In practice, double Blair Latrines smell if the pit is not properly divided.

In the double compartment version it is possible to build pits which are round or square, with slabs which are round or square and structures which are round or square. The main description in this chapter applies to a double square pit, fitted with square cover slabs and a double square structure.

The construction

Tools required

Pick, shovel, bucket, wooden and steel floats, tape measure, rope, trowel, spirit level, string and hammer.

Building materials required

1. Cement 10 pockets
2. Bricks approx. 3000
3. River sand (approx. $1 m^3$)
4. Pit sand (approx. $3 m^3$)
5. Gravel (approx. $1/4 m^3$)
6. Reinforcing wire (50 m × 3 mm)
7. Chicken wire (4 m × 2 m)
8. Flyscreen (2)

A double compartment tank version of the Blair Latrine serves a Police station in a peri-urban settlement.

Siting

This should be downhill and at least 30 m away from a well. It should be sited downwind of the house with the opening facing towards the house. It should be built on slightly raised ground so that rainwater will drain away. It should also be placed where there are no trees to disturb the airflow across the pipe.

The pit

Once the pit has been correctly sited, it is marked ready for digging. It should be 1.9 m wide and 3.5 m long. The pit is dug 3–4 metres deep, with straight walls. The pit is lined from the bottom with cement

mortared brickwork to form a 4.5 inch (112 mm) wall. A 112 mm brick dividing wall should also be built up from the bottom, so that the two pits are entirely separate. The mixture for this mortar is 8 parts pit sand to 1 part cement. The pit dividing wall should not have spacing, and should be airtight from top to bottom.

The collar

A rectangular line of bricks 225 mm (9 inch) wide is laid in mortar around the head of the pit to form a collar. This is levelled off to accept the cover slabs.

The coverslabs

These are made to the measurements shown in the diagram. Each slab is 1.75 m × 1.9 m and 75 mm thick and will require about one pocket of cement. If gravel is available the mixture should be 3 parts gravel, 2 parts sharp washed river sand and 1 part cement. If gravel is not available the mixture should be 5 parts river sand and 1 part cement.

The slab is constructed by arranging a rectangle of bricks on the ground as shown in the diagram. It is best that a piece of plastic or paper is laid down first. Two holes will be made in each slab, one for

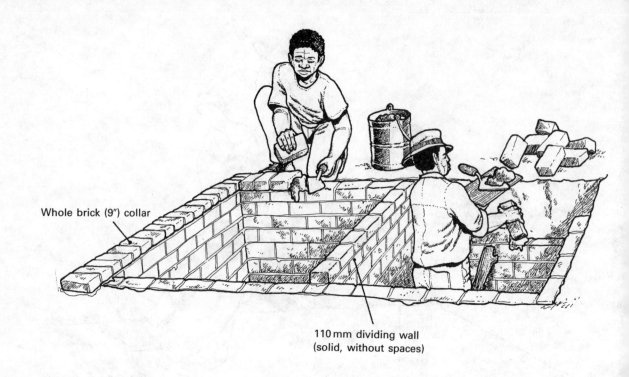

Whole brick (9") collar

110 mm dividing wall (solid, without spaces)

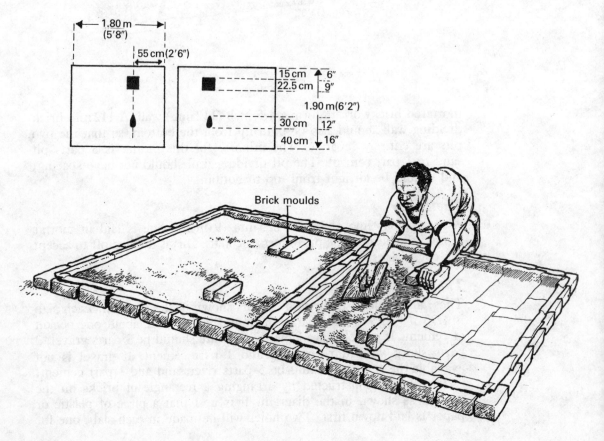

Brick moulds

the vent pipe and one for the squat hole. These are made by placing bricks or suitably shaped templates within the slab mould as shown in the diagram. The size of the vent hole will depend on the type of pipe chosen. If a brick vent is used, two bricks can be laid down to leave a hole measuring 225 mm × 225 mm in the slab. If a PVC pipe is chosen, this can be used to make a suitable-sized hole. Suitably shaped bricks can be used to make the shape for the squat hole which should be about 150 mm wide and 300 mm long.

Once the mould is set up, some 3 mm reinforcing wire is cut and laid out over the two slabs in the form of a grid with 100 mm spaces. Once the wires have been cut to length, they are laid on one side and the concrete mixture is made up. Half the mixture is laid in the mould and trowelled flat. The reinforcing wires are placed in position and the remaining concrete added and trowelled flat. The slab is covered with grass or sand and left for at least five days to cure. It should be kept wet at all times.

Fitting the slabs

After curing each slab should be carefully lifted, washed and laid over the collar of the pit in a bed of cement mortar. The slabs can be rolled into position using gum poles. It is important that the two slabs meet directly over the centre dividing wall of the pit and are made airtight with cement mortar.

The superstructure

In this particular design there are no separate foundations and the superstructure stands directly on the base slab. A 112 mm (4½ inch)

wide brick wall is built up according to the dimensions shown in the diagram. The entrances should be 0.6 m wide. The arrangement for the brick pipe is also shown, each course having 6 bricks. The walls are built up to a height of 1.8 m and the vent pipes to 2.4 m. Where brick pipes are used these must each have an internal measurement of 225 × 225 mm. Internal surfaces of the brick pipes must be smoothed off. When complete the pipe must be fitted with a flyscreen, preferably of stainless steel or aluminium.

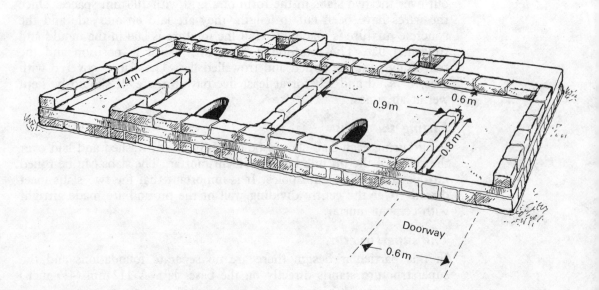

The latrine floor

This is made with a strong mixture of 3 parts river sand and 1 part cement, and laid inside the latrine, so that the floor slopes down from a step made at the entrance to the squat hole. This is finished with a steel float.

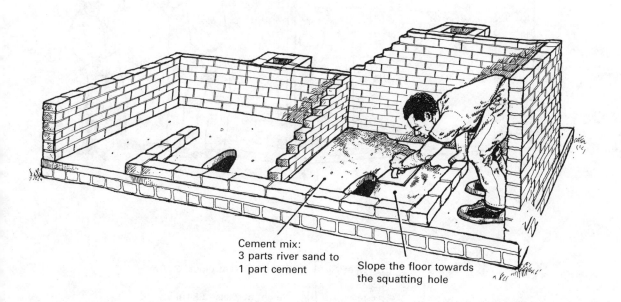

Cement mix: 3 parts river sand to 1 part cement

Slope the floor towards the squatting hole

The roof

The roof slabs are made in four pieces from ferrocement. The roof is made by levelling off a piece of ground on which the roof is to be cast. If plastic is available lay it down first where the roof is to be made as this will make lifting the roof easier. The sizes of the slabs are indicated on the diagram and are marked on the ground. Bricks are laid on the ground to form moulds for the separate roof pieces as shown in the diagram.

Four pieces of chicken wire (preferably 25 mm, but 40 mm will do) are cut to fit inside the roof moulds.

The concrete is now mixed in the proportion 1 part cement to 3 parts sharp well-washed river sand to make a firm slurry. The four slabs will take 2 pockets of cement. First, half of the mixture is laid within the moulds and trowelled flat. The chicken wire is then added, and the second half of the mixture is laid and trowelled flat. The roof should be about 25 mm to 30 mm thick.

Once cast, the roof slabs must be left to cure for one week. They should be covered with grass or sand and kept wet. After curing the bricks should be removed and each section carefully lifted, washed and added to the structure, being laid on a bed of cement mortar.

The roof can also be made of corrugated iron or asbestos sheets if this is available and affordable.

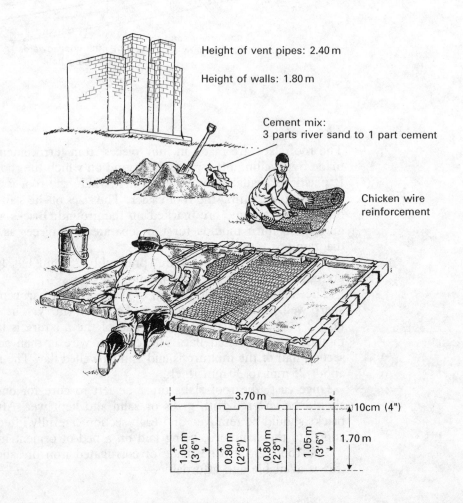

Height of vent pipes: 2.40 m

Height of walls: 1.80 m

Cement mix:
3 parts river sand to 1 part cement

Chicken wire reinforcement

Finishing off
Make sure the flyscreen is added to the top of the vent pipe. This should be mortared in position on the top layer of bricks. All inside walls of the structure should be plastered with cement mortar. Soil should be built up around the latrine so that the ground is raised slightly, thus preventing excessive erosion during the rains. Planting grass in this soil helps stabilisation. Latrines can also be painted to improve appearance.

Note
Double compartment Blair Latrines can also be made over round pits fitted with round cover slabs, as illustrated.

The multicompartment version

The Blair Latrine functions well, whether it is built as a single, double or multicompartment unit. The multicompartment unit described here was designed for schools, as it is more economical in terms of land use and bricks, than building the same number of individual units.

The multicompartment Blair Latrine described in this chapter has ten compartments, which is suitable for 250 pupils. Almost the entire structure is built with bricks. In a school setting it is advisable to build a urinal for use by the boys, and connect this through an underground pipe to one of the end pits of the main latrine block. This will be described in the next section.

Blair latrines operate effectively when one squatting hole is fitted over one pit in combination with one vent pipe. If two squat holes are fitted over one pit, with either one or two vent pipes, at least one of the latrines will smell and attract flies. This is because the pipe which draws air through the system remains effective if the air is drawn in through one squatting hole. If two squatting holes are made over one pit, air will pass down one and up the other, and the pipe may have little influence over the air currents. Since Blair Latrines are designed to be odour free, it is important that each compartment of a multicompartment unit is built over a separate pit, and that each pit has its own vent pipe.

Double compartment version over round pit.

(Measurements in metres).

(Note: This illustration shows a PVC vent pipe placed on the same side as the entrance. The pipe can also be built with bricks. Builders often prefer to place the pipe and the entrance on the same side of the structure, as the slab measurements are simpler to remember.

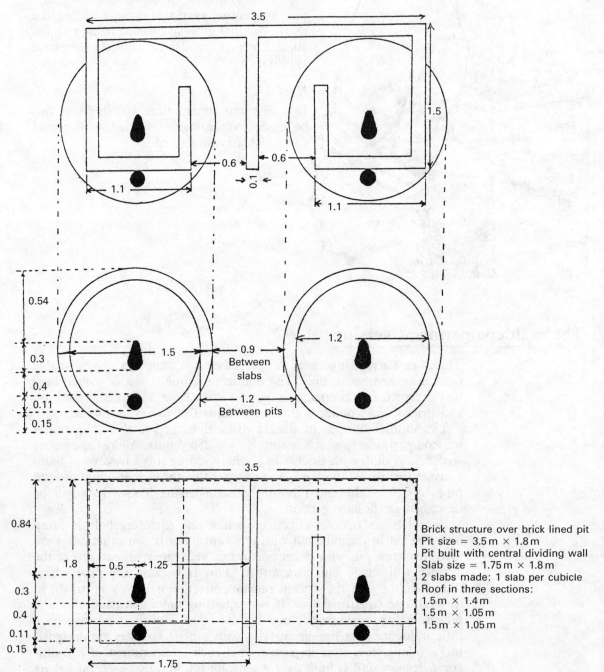

Blair Latrine.

Brick structure over brick lined pit
Pit size = 3.5 m × 1.8 m
Pit built with central dividing wall
Slab size = 1.75 m × 1.8 m
2 slabs made: 1 slab per cubicle
Roof in three sections:
1.5 m × 1.4 m
1.5 m × 1.05 m
1.5 m × 1.05 m

Two systems are possible for multicompartment latrines. One uses a large subdivided brick lined pit, the other a series of separate tubular brick lined pits. The former system is preferable and is described in this chapter. Plastered pits are easier and cheaper to build, but are less stable than brick lined pits. If water is freely available and the latrines are washed down daily it may be preferable to fit tanks under the latrines, rather than pits. These can be pumped out from time to time, and can also be fitted with flush toilets at a later date. These features make the latrine permanent and upgradeable.

The advantage of having a number of separate cubicles is that one class or grade can be assigned the task of keeping one particular cubicle clean. One problem school latrines have in common with all communal installations is that they can suffer from a lack of maintenance. If the latrines are not washed down regularly they become smelly, mainly due to urine which can seep into the concrete floor.

The multi-compartment Blair Latrine at a school.

The construction

Tools required
Picks, shovels, buckets, wooden and steel floats, tape measure, rope, trowels, spirit level, string and hammer.

Building materials
1. Cement approx. 50 pockets
2. Bricks approx. 10,000
3. River sand (approx. 5 m^3)
4. Pit sand (approx. 15 m^3)
5. Gravel (approx. 1½ m^3)
6. Reinforcing wire (250 m × 3 mm)
7. Chicken wire (20 m × 2 m)
8. Flyscreen (10)

Siting
This must be decided at the school. It should be conveniently placed downwind of the main school blocks, with the openings facing towards the school buildings. The soil should be firm and slightly raised. It should not be located near a well or a borehole (over 30 m) and away from trees which may interfere with the ventilation in the pipes.

The pit
Once the latrine has been correctly sited, a pit should be dug 14 metres long, 2 metres wide and 4 metres deep. The sides of the pit should be dug straight. Next the pit is lined with cement mortared brickwork from the bottom to one course above ground level. This should be 112 mm (4½ inch) thick. The pit dividing walls are also built up to the same height from the bottom. The bottom of the pit is not mortared, and is left for seepage. Measuring from the outside of the brickwork, the total length of the brick lining is 13.6 metres and the total width 1.8 metres.

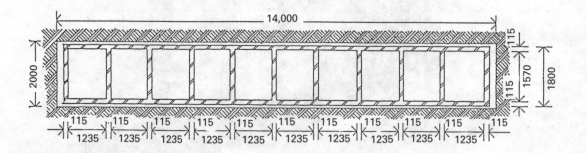

The coverslabs
Two types of slab are required, left-handed and right-handed. These are made according to the dimensions shown in the diagram and are 75 mm thick. Each slab requires about one pocket of cement. If gravel is available the mixture should be 3 parts gravel, 2 parts river sand and 1 part cement. If gravel is not available, the mixture should be 5 parts well-washed sharp river sand and 1 part cement.

The slabs can be constructed by suitably arranging lines of bricks on a flattened piece of ground. These act as a moulds for the slabs. It is best if plastic sheet can be laid down first. Two holes are made in each slab, one for the squat hole and one for the vent pipe. Multicompartment latrines can be made with brick pipes or PVC pipes. The diagrams show the slabs designed for the 110 mm PVC pipe. If brick pipes are chosen the hole in the slab should be 225 mm × 225 mm. The squat holes should be approximately 300 mm long and 150 mm wide.

Once the moulds are set up, lengths of 3 mm reinforcing wire are cut and laid in the moulds to form a grid of wires 100 mm apart. Once the wires have been cut, they are removed and laid on one side and the concrete mixture made up. Half the mixture for each slab is laid within the bricks, and trowelled flat. The reinforcing wires are then fitted, and the remaining concrete added and levelled off. All the slabs are built in the same way, making sure that five are right-handed and five are left-handed. The slabs are covered with grass or sand and left to cure for at least five days. They should be kept wet during this period.

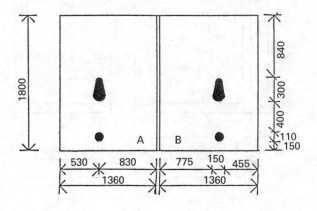

Left and right cover slabs.

Fitting the slabs
After curing, the ten slabs are fitted neatly over the pit. Care must be taken to lay the right hand side and left hand side correctly, so that a single entrance can be made for two cubicles. Each slab should fit over one pit with the edges of the slabs meeting over the dividing wall. The slabs are mortared in position to make them airtight.

The superstructure
This is built up with 112 mm (4½ inch) brickwork, following the dimensions shown in the diagram. In this design, a single entrance serves two cubicles, which economises on the use of bricks. If a brick pipe is built this forms part of the superstructure. Each course in the pipe must be built with six bricks to give an internal dimension of 225 mm × 225 mm. Internal surfaces of the brick pipe should be smooth. The pipe should be built at least half a metre above the roof level.

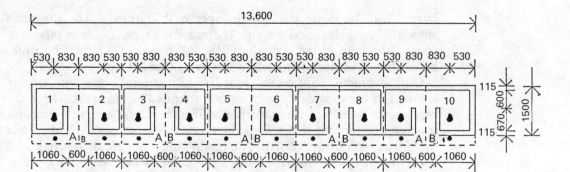

Dimensions of the full structure.

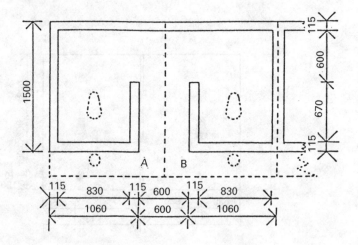

Dimensions of a cubicle.

The latrine floor
This should be laid down within each cubicle along a slope from a step made with a single line of bricks at the entrance to the squat hole. The concrete used for this job should be made with 3 parts river sand and 1 part cement. It should be steel floated.

The ventilation pipes
The diagrams show the use of 110 mm PVC pipes, but brick pipes may be preferred in the rural setting. These are made as part of the superstructure as shown in the diagram. Ensure that the vent pipe is fitted with a screen, preferably of stainless steel or aluminium.

The roof
This is an important part of the structure. The roof is most economically made with ferrocement using a mixture of 3 parts river sand and 1 part cement, the reinforcing being chicken wire. Each twin cubicle is fitted

with a roof made in three parts as shown in the diagram. As in the case of the base slabs, bricks can be used as moulds. The roofs are cast on the ground within these brick moulds to the required dimensions.

Suitably-sized pieces of chicken wire are cut so that they fit neatly within each mould. 25 mm chicken wire is best but 40 mm chicken wire can be used with care.

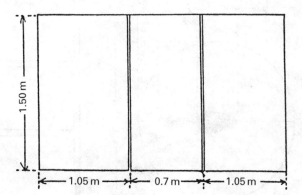

Ferrocement roof cast in three sections.

The concrete is mixed to form a thick slurry, and this is added to one roof mould at a time to make a depth of about 12 mm (½ inch). This is levelled out and the chicken wire is added. A second layer of slurry is now added and trowelled flat to make a final thickness of 25 mm to 30 mm. Each section of the roof is made in the same way. They are covered with grass or sand and left to cure for one week. They should be kept wet all through this period.

After the curing period, each roof is carefully raised, washed and fitted on to the structure in a bed of mortar as shown in the diagram.

Finishing off
The inside walls of the structure are plastered, so they can be washed down more easily. To improve the appearance of the whole unit, the outside walls can be plastered and painted. Soil should be built up around the base of the latrine to prevent erosion during the rains.

Maintenance
To maintain an odour-free school latrine it is essential that each cubicle is washed down each day with water. If this is not carried out, the cubicles will begin to smell of urine. Ideally one class should be assigned the task of maintaining one cubicle.

Note
Multicompartment Blair Latrines can also be made with two or more cubicles designed as a urinal. This design is described in the section on urinals.

The multicompartment latrine can also be made over round pits fitted with round cover slabs. This version is shown in the following diagram.

Blair multicompartment Latrine.

Roof hatched area indicates that roof should be made in two pieces

Brick superstructure

540 mm · 300 mm · 400 mm · 110 mm · 150 mm

1500 mm

Circular slab

0.9 m
0.8 m
1.1 m
1.7 m
1.5 m
0.6 m

Brick lined and subdivided trench, depth at least 3 metres

Position of structure

Circular pit 1.2 m diameter

1.2 m
1.2 m
0.2 m
0.5 m

Rectangular slab 1.8 m × 1.7 m

1.7 m
1.25 m

Slab supported by dividing walls

1.8 m

This design uses 10 squat holes and is suitable for 250 pupils.

A urinal for school latrines

Much of the waste passing into a school latrine consists of urine and it is advisable either to build a urinal as part of the multicompartment unit or build a separate urinal and drain the waste either into a soakaway or into one of the pits of the multicompartment unit. If the second option is chosen, the connection can be made underground with thick walled PVC piping.

In both cases the urinal is an elongated single Blair Latrine structure, designed so that the walls are finished with smooth and very hard mortar. Urine passing from the walls is led through a channel either directly into the pit through a hole in the base slab, or in the case of the separate urinal, through a pipe leading to a soakaway or into one of the pits of the existing Blair Latrine. The siting of the separate urinal is the same as for the latrine itself.

Urinal within multicompartment unit

In this case five cubicles with single squat holes are retained for the girls and three for the boys with the remaining two cubicles being combined into a single urinal. There should be a single pit under the urinal to accept urine. It is important therefore to decide whether a urinal is to be incorporated into the multicompartment structure from the beginning.

Base slabs

Where square slabs are used to cover a subdivided rectangular pit, two will be used under the urinal section. These will obviously not have the normal squat holes cast into them. One of these two special slabs will have no holes cast in it while the second will have a hole for the vent pipe and a 75 mm diameter hole for urine drainage cast into it.

Where round slabs are used over round pits, only one pit is required for the urinal and the second cubicle can be built over a foundation which does not cover a pit. This is illustrated in the diagram on page 328.

Urinal cubicle

This is constructed in the same way as two normal cubicles, but with the dividing wall missing, so that one large cubicle is formed. One base slab will be without holes, the other will have one hole for the vent and another for urine drainage, positioned as shown at the base of the wall. The cubicle is built up to normal height and the internal walls plastered with a hard steel floated cement mortar for one metre above floor level.

Urinal floor and channel

A series of bricks is laid on edge 150 mm away from the wall and mortared in position side by side as shown in the diagram. The channel is made by backfilling the space between the wall and the line of bricks with strong cement mortar so that it slopes down from the level of the

This urinal system was designed by Johnston Chinyanga of Save the Children's Fund UK.

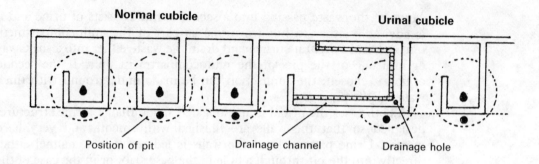

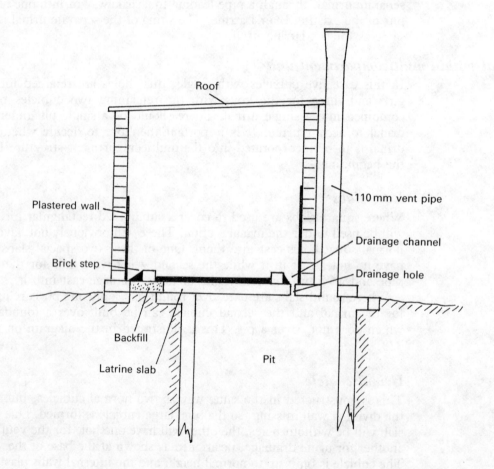

Urinal in multicompartment latrine.

brick to the base of the wall. The channel is built at the base of three of the walls of the urinal as shown. The channel is sloped so that all the urine collected drains through a drainage hole into the pit beneath.

The urine floor is built up on the base slab between the channels in strong cement mortar, and steel floated.

Construction of separate urinal

It is assumed that a multicompartment latrine has already been built. If not, a soakaway should be built to accept the urine from the urinal.

Foundations and walls

The first stage is to clear and mark the ground and lay a brick foundation for the walls of the urinal. The foundation trench should be 300 mm wide and 150 mm deep. The size and shape of the foundation and brick walls of the structure are shown in the diagram. The structure should be 5 m long and 2 m wide, with a doorless entrance, as in the single compartment Blair Latrine. The foundation itself should be made of bricks, laid side by side in mortar, 225 mm wide and 250 mm deep (3 courses). This will bring it to about one course above ground level. The 112 mm (4½ inch) brick walls should be built up from the foundations to a height of 1.8 m.

Urinal base slab

The base slab of the urinal is now laid down within the foundations. This should be made with concrete (4 parts gravel, 2 parts river sand and 1 part cement), reinforced with 3 mm wire and made 75 mm thick. This is flattened with a wooden float and allowed to cure for one day.

Urinal floor

This is laid on a slight slope, 50 mm to 75 mm deep, with concrete (mixture as above, or if gravel is not available 5 parts sharp river sand to 1 part cement). It is laid over the base slab as shown in the diagram. It is necessary to lay the floor so that it slopes and runs into a channel. The channel is formed at the base of the wall and at the lower end of the sloping floor by laying down a series of bricks at the base of the wall, as shown in the diagram, before the urinal floor is laid down. This floor is then built up against the bricks. When the urinal floor has set, the bricks can be removed to leave a channel.

Fitting the drainage pipe

Before the final plaster work is laid over the working surfaces of the urinal, it is necessary to lay the drainage pipe between the urinal channel and the end pit of the Blair Multicompartment Latrine (or a soakaway). PVC pipe can be used for this purpose, with 50 mm class 16 being the best. A 50 mm PVC bend will also be required. The length of pipe will depend on the distance between the urinal and the soakaway or Blair Latrine.

A section of the brick foundation is excavated so that the 50 mm PVC bend attached to the length of 50 mm PVC pipe can be introduced through a gap left in the foundation. The bend should be positioned so that it can drain urine from one end of the channel into the pipe that runs through an excavated trench to the soakaway/pit along a slight slope. This is shown in the diagram. The excavated section of foundation is built up again in concrete around the bend and drainage pipe.

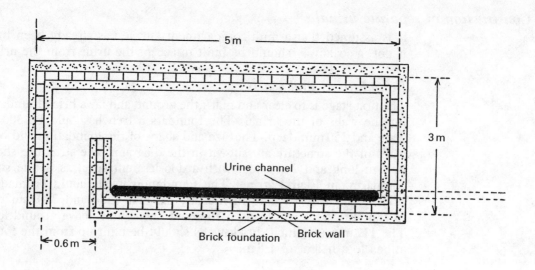
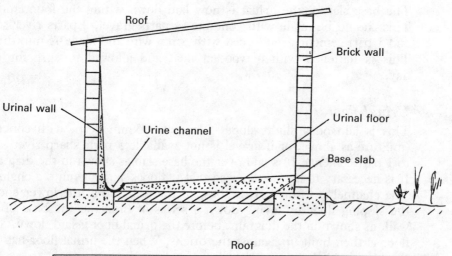
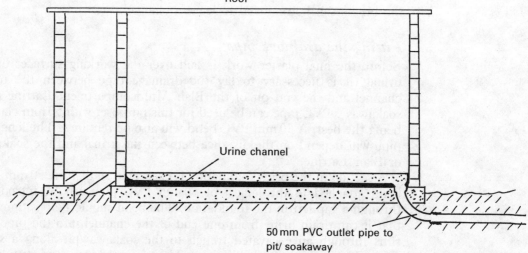

School urinal.

The drainage pipe passes through an excavated trench which leads to the pit lining wall at the end of the latrine. A hole is knocked through the wall at the appropriate point and the pipe passed through a short distance. The wall is cemented up again around the pipe. The trench is refilled and levelled.

If a soakaway is used for drainage this can be made by digging a pit 1.5 m across and 2 m deep and lining it with bricks, which should be spaced apart at the bottom of the pit. A brick collar is made at the top of the pit and covered with a concrete cover slab. The urinal drainage pipe can be led into this pit. If drainage conditions are poor, the soakaway may require enlarging.

Plaster work

It is now necessary to plaster a hard working surface of cement mortar on the urinal wall, the urinal channel and the urinal floor. This is an important part of the unit and should be steel floated with a river sand/cement mix of 3:1.

First plaster the urinal wall down from a height of 1 metre. This can slope slightly towards the channel. The channel itself can be plastered to leave a smooth internal surface, and once again sloped slightly so that urine will drain towards the drainage pipe. The urinal floor is also plastered smooth and this will also slope slightly towards the channel.

The remaining internal surfaces can also be plastered with normal mortar, 5 parts pit sand to 1 part cement.

Roof

A roof is not essential for a urinal, but is useful during the rains and aids the appearance of the structure. This can be made by casting a series of ferrocement slabs using a river sand cement mix of 3:1 reinforced with 25 mm chicken wire. Alternatively asbestos or iron roofs can be used.

Finishing off

The area around the urinal can be built up with soil, and the trench levelled off. The finished structure can be plastered and painted. It is advisable to place a stainless steel sieve over the channel drainage hole to avoid blockage by sticks and stones.

Maintenance

The urinal will require washing down regularly with water. It may be necessary to rod the drainage channel from time to time to clear it of sticks and sand which may accumulate in the pipe.

The tank and soakaway version

Most Blair Latrines are built over pits, which is ideally suited for most rural development programmes, where space permits the construction of a second or third latrine on one family stand or for a new pit to be dug when the old one fills up. The family unit being promoted in rural areas should last a family for fifteen to twenty years if the facility is regularly washed down and paper is used for anal cleansing. If other

objects are added to the pit the useful life is reduced. When full, the old pit is normally abandoned and a new pit dug. Very often components from the original latrine — slab, roof, bricks and vent pipe — can be re-used on a new structure. Ideally each family should have two units, which greatly extends the life of each pit. In recent programmes, the first unit is built with a subsidy from Government or a donor. The second unit should be built entirely by the family themselves.

The situation described above does not apply in all situations however. In peri-urban settings, for instance, space may be limited on each plot. In this case it is necessary to build a latrine which is not only permanent, in every sense of the word, but also has features which make it upgradeable into a fully waterborne flush unit. In cases like this a normal pit-type Blair Latrine is inadequate since the pit will become flooded when the upgrading process takes place. In addition, dry pits are not easy to empty, and require specialised equipment. The contents of a permanently sited latrine should be easy to desludge. The Tank and Soakaway version of the Blair Latrine has been designed to cope with these particular situations. This type of Blair Latrine was first placed on trial in 1976. It was designed for three basic reasons:

1. To increase life expectancy of the pit/tank.
2. To make desludging easier than for a pit.
3. To increase the capacity for accepting waste water and thus make upgrading to a waterborne/shower system possible in the future.

Operation

It is well known that excreta is digested more rapidly in wet conditions than in dry ones. Digestion is particularly efficient in the 'septic tank.' In the normal septic tank, waste matter is carried in pipes from the toilet into the middle (effluent) layer of the tank. Fresh waste rises to digest in the upper (crust) layer and then falls to the bed of the tank where it becomes known as sludge. Excess liquids are fed away through an effluent discharge pipe to a soakaway. In a normal septic tank, wastes are added into the effluent layer, whereas in the version described here, raw sewage is added to the top of the crust layer. Experience has shown, however, that a fluid effluent layer and a sludge layer are formed and that the same microbiological processes take place in both systems.

The advantage of the tank version of the Blair Latrine is that the digestion of waste proceeds more efficiently than in a pit and that the contents are mobile enough to be pumped out, after rodding, with a conventional tanker. This would not be possible with the contents of a pit. Because the system also incorporates a soakaway for the waste effluents, it can absorb a greater volume of waste liquids. This makes further upgrading of the latrine possible so that either a shower unit or a flush toilet can be added later.

In some situations the amount of water added to a pit type Blair Latrine can exceed the rate of seepage from the pit, and the water level rises quite high in the pit. This happens normally in areas of poor drainage, where the soils are highly compact or have a high content of clay or rock. In such cases the addition of an effluent discharge pipe which taps off the fluids at a certain level, and carries them to a larger soakaway, works much better. The degree of plugging at the base of a soakaway by sludge carried in effluent is far less than the plugging at the bottom of a pit with raw excreta. It is also possible to carry the effluent away from the immediate area, where drainage may be poor and allow it to soak away in more suitable ground. Similarly it is possible to tap off the effluent discharge from several tanks and feed them into a common soakaway or a waste stabilisation pond.

The design of the tank and piping in this version is different from that of conventional practice. The tank is 3 m deep and 1.2 m in diameter, the liquid depth being 2.25 m. The tee-shaped effluent discharge pipe rises 50 cm above the liquid layer and 50 cm below it. The effluent discharge pipe is made with thick-walled PVC (63 mm class 16 PVC). The tee section is built directly below the vent pipe, to facilitate inspection and rodding if necessary. The horizontal component of the discharge pipe has been placed 75 cm down in the tank to accommodate for a liberal build up of sludge. The considerable depth of the tank has been designed to accept a certain quantity of non faecal matter including sand. Since there is no restricting water seal which has the effect of reducing the range and number of extraneous items that might find their way into the tank, it has been assumed that some solid objects might find their way into the tank. 1.75 metres of tank depth is available for sludge accumulation. The accumulation rate of sludge has been measured in five year old tanks of this type used by families. The rate of accumulation of sludge at the bottom of the tank is approximately $0.013 \, m^3$ per person per year. The accumulation rate depends on many factors including the age of the facility. Sludge compacts as it becomes older.

The total waste accumulation capacity of a 1.2 m diameter tank would be approximately $1.97 \, m^3$. Thus for a family of ten persons there is adequate capacity for sludge accumulation for over ten years. Larger capacity tanks having a diameter of 1.5 metres have a total waste accumulation capacity of $3.09 \, m^3$. This volume of sludge would accumulate from a family of ten in fifteen to twenty years. These figures are approximate and calculated from tanks in their first five years of life.

The volume of waste formed in the first year of operation continues to decrease in each successive year, partly by further decomposition, partly by compaction and partly because some solids are taken away in the effluent into the soakaway, where further decomposition takes place. By comparison a 3 m deep 1.2 m diameter pit would last a family of ten for about twelve years, but at this stage it would be full, and quite solid and difficult to empty. The exact accumulation rate varies from one family to another.

Although the frequency of sludge removal from tanks of this type has not been exactly worked out, it may lie between five and ten years, depending on family size. In small families the tank may run for decades without requiring desludging. In larger families emptying at five year intervals may be required. Conventional tankers can desludge the unit, if the sludge is rodded and stirred first. In a peri-urban settlement served by this system a single small tanker might effectively empty the tanks at a rate of perhaps three to five units per day every working day on a rotation basis. Thus approximately 500 units might be desludged every year. If each unit required desludging every five years, the tanker might serve approximately 2500 units.

The construction

A series of diagrams illustrating the construction are included in this chapter. Constructional details for the superstructure can be found in the chapter describing the single Blair Latrine. Two tank and soakaway sizes are recommended. Smaller families of up to ten persons should have a tank diameter of 1.2 m and a soakaway diameter of 2.0 m. Larger families should have a tank diameter of 1.5 m and a soakaway diameter of 2.4 m. Communal installations may require larger tanks and soakaways depending on the anticipated number of users. The following description applies to families of up to ten persons.

Building materials required
1. Cement 10 pockets
2. Bricks 1500
3. River sand 1 m^3
4. Pit sand 3 m^3
5. Gravel (12 mm chips) 1.5 m^3
6. Reinforcing wire 50 m × 3 mm
7. Chicken wire 1.8 m × 2 m
8. Vent pipe 110 mm stainless steel screen
9. Effluent discharge pipe 63 mm class 16. 2 m horizontal, 1 m vertical with tee
10. PVC soakaway inspection pipe 110 mm class 6 (0.5 m)

Excavations

Dig 2 holes 1 m apart. The tank hole is 1.5 m in diameter and 3 m deep. The soakaway hole is 2.0 m in diameter and 1.5 m deep.

Dig a trench 75 mm deep between the holes.

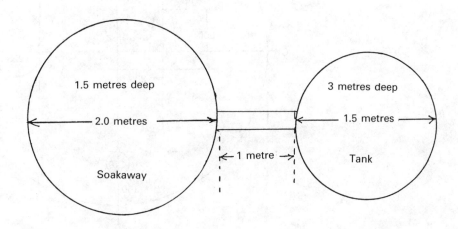

Casting concrete slabs

Cast 2 concrete slabs, one for the soakaway lid and one for the latrine base slab. Both are made 75 mm thick as shown in diagrams with a mixture of 4 parts stone, 2 parts river sand and 1 part cement. Use 3 mm wire for reinforcing in a grid pattern at 100 mm intervals. With care, the soakaway slab can be made 50 mm thick. Use templates to make the holes required in the slabs.

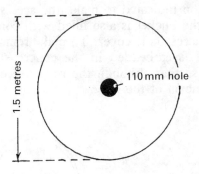 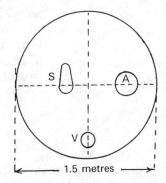

For slab measurements see next diagram

S — Squat hole
A — Tank access hole (fitted with cover)
V — Vent pipe hole

337

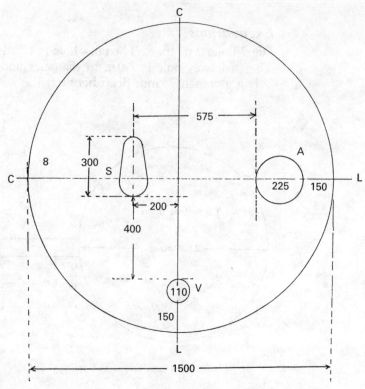

Blair Latrine (tank version).
Slab measurements in mm.

Method of making tank access hole and cover

The tank access hole is situated outside the structure for ease of rodding and removing the contents with a tanker and hosepipe.

The access hole is 225 mm in diameter and can be made by using a mould made from the upper 125 mm (5 inch) half of a 10 litre plastic bucket. When the slab is being made, a piece of 110 mm PVC pipe serves as a mould for the vent pipe hole, and bricks (or shaped mould) can be used to make the squat hole. The upper section of the bucket is placed in the position indicated to make the access hole. At the same time the inside of the bucket is also filled with concrete about 80 mm deep, to make the access hole cover. To make removal easier, a length of 16 mm steel rod is embedded in the concrete and recessed. The concrete is shaped so that although the rod is recessed a hooked tool can be used for removal of the cover.

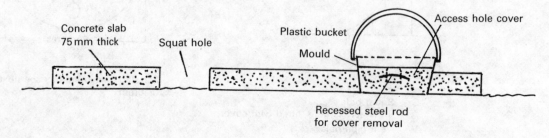

Tank and soakaway

Add a 75 mm thick layer of concrete to the base of the tank pit. Reinforce with 3 mm wire and leave it to set. Build up the walls with brickwork 112 mm thick with mortar to the level of the ground to leave an internal diameter of 1.2 m. Plaster the inside of the tank (5 parts pit sand, 1 part cement). Add a 225 mm wide ring of bricks at the head of the tank to make a collar.

Build up a tube of unmortared brickwork within the soakaway pit, 1 m high and 1.5 m external diameter. Use fired bricks.

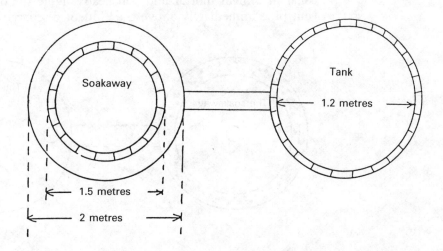

Add a PVC soakaway pipe between the tank and the soakaway as shown, knocking a hole in the brick wall of the tank, and replastering around the pipe. Pass this pipe through the brickwork of the soakaway.

Fill the space between the bricks and the earth wall of the soakaway with 12 mm granite chips or gravel to form the soakaway.

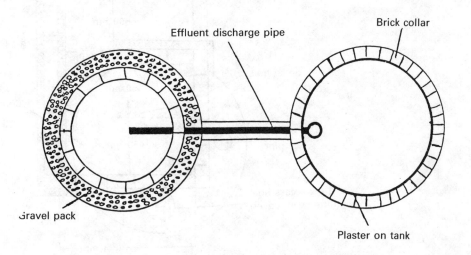

Add a 1.5 m diameter slab on top of the brickwork in the soakaway hole. Place plastic sheets over the gravel in the soakaway. Add a soakaway inspection pipe to the soakaway slab. Backfill the soakaway and trench with soil and ram level.

Add a concrete protective apron around the upper end of the soakaway inspection pipe 300 mm in diameter. Cast a suitable concrete lid to cover and seal the pipe. Mortar this in position over the inspection pipe.

Cement the vertical section of the effluent discharge pipe to the horizontal section within the tank. Special PVC cement is required for this purpose. Embed the 1.5 m diameter latrine slab on to the tank collar in cement mortar and seal neatly. Note the orientation with the vent pipe immediately above the effluent discharge pipe.

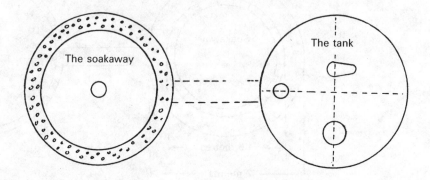

Superstructure

Lay the foundations for the structure in brickwork 225 mm wide as shown in the diagram. Note that the ventilation pipe hole is laid

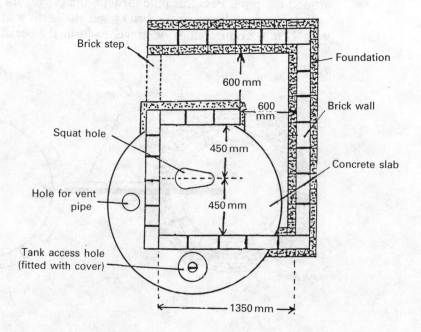

directly over the effluent discharge pipe in the tank. Build the structure walls as shown in the diagram with 20 courses of bricks. A brick step is made at the entrance and the interior floor is raised level with the slab with rubble or rammed earth.

Latrine floor

Lay the latrine floor within the structure with a mixture of 3 parts river sand and 1 part cement. This should slope down from the entrance step to the squatting hole. The internal walls of the structure are plastered with mortar.

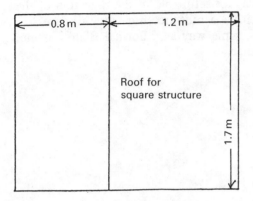

Roof

Cast the roof slabs on the ground using bricks as a mould, with the measurements shown in the diagram. The mixture used is 3 parts river sand and 1 part cement. The roof is reinforced with 25 mm chicken wire laid in the mould after half the mix has been added. The remaining half of the mix is added on top of the chicken wire and trowelled flat. This is cured for one week, being kept wet at all times. It is then carefully lifted, cleaned and mounted on the structure in a bed of mortar.

The ventilation pipe

This is added to the vent hole in the slab and mortared in position. The top of the pipe is wired in position on to the structure.

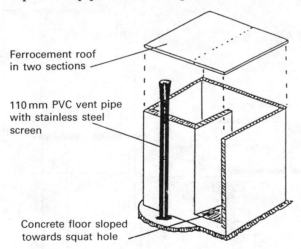

Square spiral brick Blair Latrine with 110 mm PVC vent pipe.

Normal operation

Once the latrine has been finished it is wise to fill the tank with water at least up to the lower end of the effluent discharge pipe. From then on the tank can fill slowly until the water level reaches the tee piece in the discharge pipe and overflows into the soakaway. In the first 6 months of

341

operation it is possible that *Culicine* mosquitos will breed in the tank since the crust is not yet completely formed on the surface of the water. Although some of these mosquitos will be trapped in the pipe, many will escape through the squat hole and become a nuisance. It is possible to overcome the mosquito problem by adding a half kilogramme layer of 3 mm diameter polystyrene balls to the surface of the water in the tank. Once the crust has built up no more mosquitos can breed in the tank, if polystyrene balls cannot be obtained.

In some tanks small insects known as *psychodas* or tank flies can develop. They rest on the side walls inside the latrine and are common with many septic tanks. They play a useful role in the digestion of the sewage. Fortunately they are neither a nuisance nor carry disease.

The latrine will operate in the same way as a normal Blair Latrine, and requires the same maintenance.

Blair Latrine.
(Upgradeable tank and soakaway version).

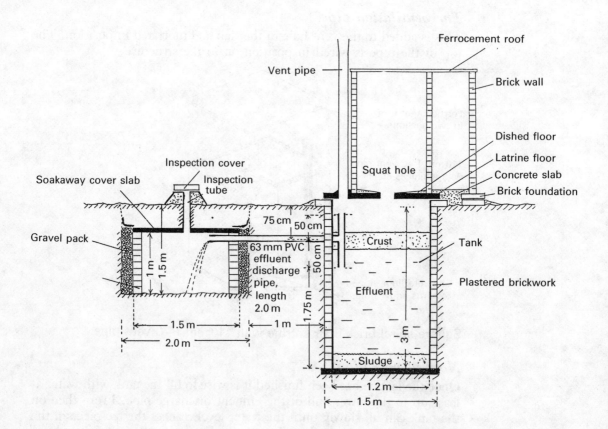

Construction of double compartment Blair Latrine (tank version)

Note: the dimensions shown beneath are for a double compartment communal unit. In this version the tank is larger than a family unit and has a final internal diameter of 1.5 m. Family unit tanks have an internal diameter of 1.2 m. The soakaway size is shown here as 2 m in diameter. The size of the soakaway will vary depending on ground conditions.

Stage 1. Tank and soakaway holes
Dig 4 holes as shown in the diagram. The two holes for the twin tanks are 3 m deep and 1.8 m in diameter. The twin soakaway holes are 1.5 m deep and at least 2.0 m in diameter. The walls must be vertical.

Dig 2 trenches between the tank hole and the soakaway hole as shown to a depth of 75 cm. The tank and soakaway holes are 1.0 m apart.

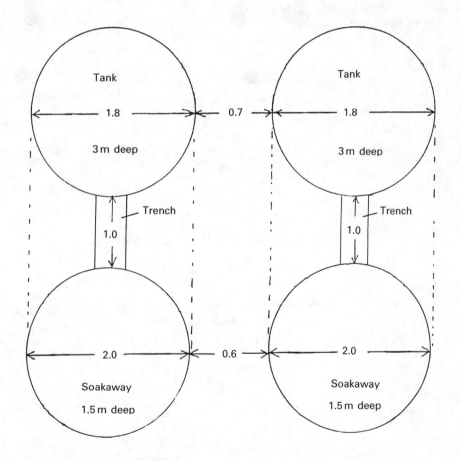

Stage 2. Concrete slabs
Make 4 concrete slabs 75 mm thick with 3 mm reinforcing wire at 100 mm spaces. The mixture is 3 parts gravel, 2 parts river sand and 1 part cement. If gravel is not available 5 parts clean river sand and 1 part cement will do.

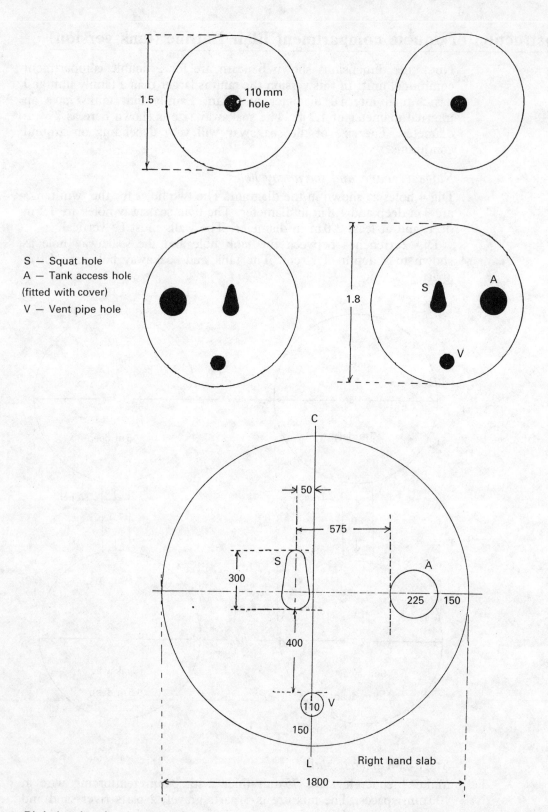

Blair Latrine (tank version).
Slab measurements in mm.

Make 2 larger slabs for the latrine base and 2 smaller slabs for the soakaway lid. Note that the squat hole dimensions are 300 mm × 150 mm. The vent pipe hole is 110 mm diameter for PVC pipe. The soakaway inspection pipe hole is 110 mm diameter. The concrete should be allowed to cure for one week.

Method of making tank access hole and cover

The tank access hole is situated outside the structure for ease of rodding and removing the contents with a tanker and hosepipe.

The access hole is 225 mm in diameter and can be made by using a mould made from the upper 125 mm (5 inch) half of a 10 litre plastic bucket. When the slab is being made, a piece of 110 mm PVC pipe serves as a mould for the vent pipe hole, and bricks (or a shaped mould) can be used to make the squat hole. The upper section of the bucket is placed in the position indicated to make the access hole. At the same time the inside of the bucket is also filled with concrete about 80 mm deep, to make the access hole cover. To make removal easier, a length of 16 mm steel rod is embedded in the concrete and recessed. The concrete is shaped so that although the rod is recessed a hooked tool can be used for removal of the cover.

Stage 3. Tank and soakaway construction

Add 75 mm depth of concrete to the bottom of each tank to make the base. Reinforce it with 3 mm wire. Line the tank with cement mortared brickwork to ground level leaving a hole for the effluent discharge pipe (0.75 m depth). 110 mm wall thickness is adequate. Plaster the tank inside with pit sand and cement (mix 5:1).

Build up the brick tube inside the centre of the soakaway hole without mortar to a height of 1 m (external diameter 1.5 m). Fit the effluent discharge pipe between the tank and soakaway at a depth of 0.75 m. This is 2 m long. Cement the PVC tee on to the horizontal effluent pipe. Finish off the plaster around the pipe inlet in the tank. The pipe passes through the unmortared brickwork in the soakaway. The pipe is made from 63 mm class 16 PVC pipe. The upper and lower limbs of the tee are both 0.5 m long.

Add a gravel pack of 12 mm granite chips to the full depth of the brickwork in the soakaway. Add a 225 mm wide brick collar around the head of the tank.

Stage 4. Fitting slabs

Add 2 latrine slabs to the tank in a mortar bed. Locate the vent holes directly over the effluent pipe tee pieces. Add 2 soakaway slabs on top of the brickwork in the soakaways. Place plastic sheeting over the gravel pack in the soakaway.

Add 110 mm soakaway inspection pipes to the two soakaway lids and cut them off 150 mm above ground level. Fill up the trench and the remainder of the soakaway with soil and ram it firm to ground level. Add a concrete apron, 300 mm in diameter around inspection pipe and make a suitable lid in concrete to cover this.

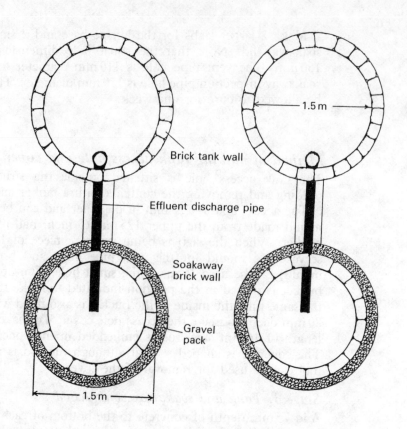

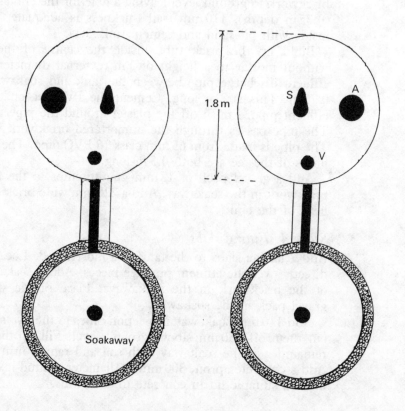

346

Stage 5. Superstructure

Lay the brick foundations for the structure 225 mm wide as shown in black in the diagram, up to the level of two slabs. Build the brick superstructure to the dimensions shown in the diagram (measurements in metres); this is 20 courses high. Bricks are laid in cement mortar.

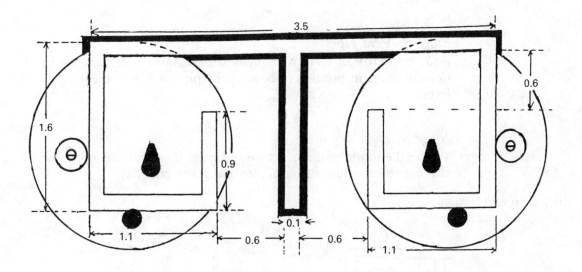

Stage 6. Roof

Cast roof slabs 25 mm thick to the measurements shown in the diagram (in metres).
Use chicken wire as reinforcement. Cement mix is 3 parts river sand and 1 part cement. Leave for 1 week to cure and keep wet. Add roof in 4 sections and cement mortar to the superstructure.

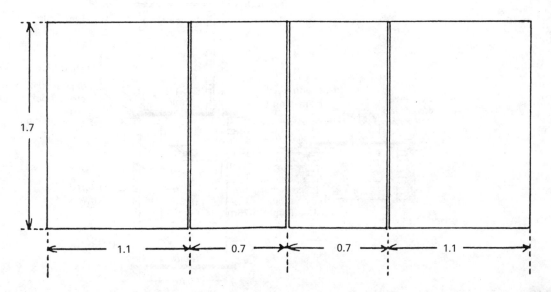

Stage 7. Latrine floor
Make a concrete floor inside the structure by adding a brick step at the entrance as shown and one course above slab level. Fill the area between the foundations with earth or half brick backfill. The concrete floor should be made from 3 parts river sand and 1 part cement. This should be sloped downwards from the entrance to the squat hole for adequate drainage.

Stage 8. Vent pipes
Add 2 × 110 mm PVC vent pipes with stainless steel screens fitted. Cement them in position at base of the pipe and wire them near roof level.

Stage 9. Finishing off
Plaster the inside of the structure and paint it to neaten it up. Level off the site surrounding the structure and plant grass.

Brick step and backfill.

Cross-section of tank version.

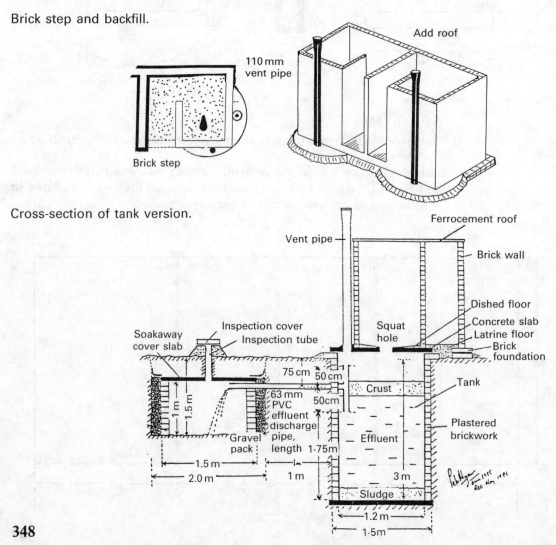

Adding a flush toilet

If a Blair Latrine has been built with a tank and soakaway it is possible to upgrade this by fitting a standard flush toilet, which can either be a squat pan or a pedestal type. It is important that the system is served with a reliable piped water supply. Water discharged from the water seal enters the tank directly, and is directed away from the effluent discharge pipe. If a squat pan is used the latrine can still be used as a bathroom. If a pedestal type is fitted there is little room left for bathing.

The new flush unit, which preferably should be of a low volume flush type, is fitted into the existing concrete slab. This will require opening the squat hole with a hammer and chisel until the unit will fit into the new opening. This should be cement mortared in position to make it stable. The floor will be raised a few centimetres to allow for the new flush unit. The floor should be suitably sloped to allow for adequate drainage, and should be steel floated with very strong cement mortar.

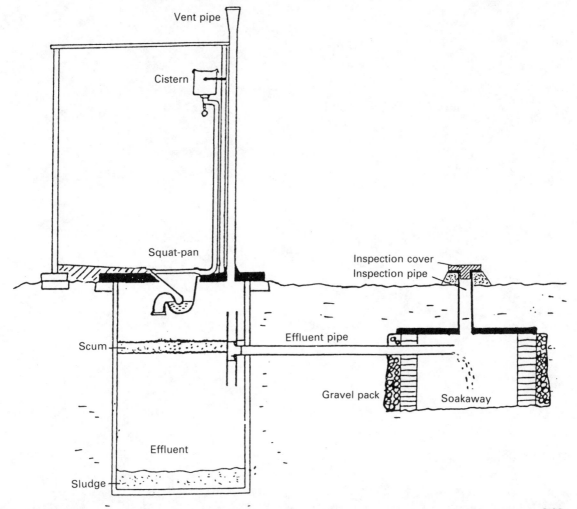

The flush unit cistern, which should be adjusted to give a five litre flush, is attached to the inside wall of the latrine superstructure and must be fed with a reliable piped water supply. This installation is best carried out by a plumber. Since flushed water only travels round a single bend, little water is required for flushing.

The external tank access hole allows for rodding and evacuation of the contents when this is necessary. Whilst in use the tank access hole cover beds neatly within the access hole itself. A special hooked tool is required to withdraw the cover when the tank is inspected or emptied.

Once the flush unit is installed the vent pipe will act as a ventilator and insect trap and can be removed for inspection of the effluent discharge pipe.

Soil conditions vary greatly and if effluent percolates slowly from the soakaway, it may require enlarging. This will not only depend on soil conditions but the amount of water added to the tank. This can only be tested on site. Extra soakaways can be built and connected either to the first soakaway or directly to the effluent discharge pipe.

Blair Latrine training programmes

Introduction

The advantages of the Blair Latrine only become apparent when it is built correctly. Given an adequate amount of building materials it is possible to build an excellent toilet that will perform correctly and give good service for a generation. Unfortunately the same materials can also be used to build a latrine which may collapse, smell and generate flies. Builders of Blair Latrines, and those who supervise construction, do their best work when they understand how the system works and when they are trained in constructional techniques. Adequate training is essential.

In the Zimbabwe programme two types of training have become common. In the first, supervisors such as Health Assistants, Health Orderlies and Community Workers, and other officials such as Health Inspectors, and Vidco Chairpersons take part in the construction of demonstration Blair Latrines. These demonstrations occur in a central place in the District or Province and are backed by theoretical training. These persons may rarely build latrines themselves, but require a good deal of background knowledge. A combination of Blair Research Bulletins and Builders Manuals serve as suitable literature.

The second type of training is designed for local builders, who are already skilled at building with bricks and mortar but require specific information about Blair Latrine construction. They are generally trained at Village or Ward level, at sites which are close to where many of the builders may live. The village is the training area.

Training of supervisors

This is undertaken at a central place in the District or Province where one latrine is built, often by the trainees (HA's, HO's, CW's etc.), under the supervision of a Training Officer (Health Inspector or Principal Health Assistant) with support from the Blair Research Laboratory. Provision of cement and other hardware comes from Provincial sources, and the finished demonstration remains as a working structure. Provision must be made for food and accommodation for the participants over a one or two week period.

Several points are stressed at these sessions. The operating principles of the Blair Latrine are described in detail with chosen examples. Constructional points are described and demonstrated in detail. These include siting and orientation of the structure, and methods of construction of pit, pit lining, collar, foundations, slab, vent pipe, superstructure and roof. Cement slabs are constructed with particular emphasis on reinforcing, and hole size and position. The importance of a good vent pipe fitted with a corrosion resistant flyscreen is also emphasised here.

Stainless steel, aluminium and PVC coated fibreglass screens are described, with preference being given to stainless steel. The importance of a sloped hygienic floor and household maintenance is also stressed. Life expectancy of the pits is also discussed.

These training excercises may take between one and two weeks for completion, depending on the amount of preparation which has taken place, e.g. digging pits and making slabs.

Training of builders

The training of builders is the most active programme, and a sequence of events takes place in preparation for this. The Provincial Health Office will have gained financial backing from Central Government often with the support of a Donor Agency, for a programme in a certain area. A supply of cement and other hardware will have been secured, and the programme is ready to begin.

Health Assistants, Health Orderlies and Community Workers from the proposed area of development are invited to the District Office on a particular day, together with a member of the Blair Research Laboratory. Together with the Health Inspector, a plan of action is prepared. Approximate dates are drawn up when the demonstration team will move into an area, Ward by Ward. By doing this each Health Assistant will know in advance when to expect the training team.

Health Officials in the chosen area now advise and liaise with Vidco Chairpersons and other community leaders and Villagers about the programme. At village level the Community Workers will be active, with the Health Assistants also being active at Village, Ward and District level.

As the time for builder training draws closer, specific sites are chosen so that about fifteen builders in a certain Ward can travel by foot or bicycle to attend the training sessions within that Ward. At that time, suitable sites are agreed upon by local people and Health Officials. Preparations are made in the form of digging pits and preparing bricks and other materials such as sand and stone. A specific date is agreed on for the activities to begin. Very often training takes place in a series of private homesteads and the family provides materials that will be used in the construction of their own latrine.

Several families may participate in any one Ward, and they may hire the builders who are being trained. The entire process of construction need not necessarily take place on one site, but may take place at many sites.

The method of lining the pit and casting the concrete slab, for instance, may be shown in one homestead, whilst methods of fitting the slab, making the foundations and structure and roof may take place at others. Trainers and builders check on the progress made in the specific area. Thus a series of demonstration toilets are built in the Ward, often at the homesteads of families who came forward quickly with a supply of bricks and sand and a fully dug pit. Finished latrines are inspected and discussed in detail by the group.

There is some variation in the exact method of training depending on the size of Wards and Villages. In some cases, one village may be chosen for the training of builders from several Wards. Builders live at home, and thus accommodation is not required for them. Local arrangements can be made for feeding etc.

Builders require specific knowledge on measurements of the slab, vent pipe and structure and advice on where specific care is required in the construction and how errors can be made, e.g. size of pipe etc. Although some theoretical instruction is given, most of the emphasis is placed on practical constructional techniques supported by Builders' Manuals. These training sessions may last one or more weeks depending on local conditions.

All members of the community are becoming more familiar with the Blair Latrine and how it works. Pupils at school are taught about it and can teach their parents at home. The importance of training at all levels is very great. A thorough knowledge of working principles and constructional techniques will assist greatly in making the National Sanitation Programme a success.

Index

Note: Numbers in italics refer to illustrations, diagrams or tables.

aeration zone 11
air currents 271–3
air locks 221
alginate 266
alum 268
aluminium sulphate 268
anal cleansing 283, 333
Anderson, Cecil 153
ant turrets 273–4, *274*
apron 91, 93, 95
    for Bucket Pump 81, 82, *82*, *104*
    for shallow wells 21
    for tubewell 59
    for upgraded well 38
aquifer 11, 13, *13*
arsenic 249
artesian depression spring 214
artesian well 14
auger 46, 47, 48

backfill 94
backwashing 265
bacteria
    counts *18*
    dilution of 74
    pathogenic 17, 18
    and quality of water *253*
    samples 42–3, *42*, *43*
    studies of 76
bacteriological data, for ground water 77
bacteriological testing kit 254
bailer valve 55
bailing 54–6
bananas
    and drainage 104
    in seepage area *240*
    in soakaway 60
    and waste water 239
base slab
    for Blair Latrine *298*
    for spiral latrine 309, *310*
baseplate 147
Bilharzia transmission 242
bitumastic 289
Blair Latrine 270–286
    basic requirements 304
    characteristics
        and capacity 283
        construction 292, *292*

essential features of *290*
and fly control 276–7
and light level 282
maintenance 286
in natural habitat *305*
and natural mechanisms 281–2
and spiral structure 278, *297*
as washroom 283–4
and cobwebs 275, 281
components
    base slab *298*
    collar 294, 304
    coverplate 283
    coverslab 294
    floor 284, 297, *299*, 304
    flush toilet 349–50
    pedestal 297
    pipes 275
    pit 293, 304
    roof 302–4
    superstructure *296*, 296
    walls 304
desludging 285
and lizards 281
and mosquitos 282
and spiders 275, 281
training in building 351, 352
types 285
    double compartment 312–21
    early *270*, *271*
    ferrocement spiral version 309–12
    flush toilet 349
    low cost *286*
    Matabeleland version 306–7
    multicompartment 321–328, *328*
    single compartment 291
    square spiral brick *314*
    tank and soakaway 333–348, *344*, *348*
    training programmes 351
    upgradeable 307–8, *308*
    upgradeable tank and soakaway version *342*
    upgradeable version 307–8
    unworkable *280*
ventilation of 270–276

in Zimbabwe *291*
Blair Pump 63, 64, 68, 109–141, 253
    assembly 118
    characteristics 112
    components
        cylinder 123, 135, 139
        footvalve 123, 136, 140
        head assembly 127
        piston valve 135, 139
        pump head 135, 139
        pushrod 123, 135, 139
        working parts *113*, 123
    description 110
    designed 109
    ease of use *114*
    for family use 110
    hand made 133
    heavy duty 137
    in hand drilled tubewell 130–1
    installation 115
    in tubewell 130
    maintenance 121–30, *122*
    Mark I 111
    Mark VI
        installed 115, *115*, *120*–1
        spare parts *124*
    types 111–12
        hand made 133
        heavy duty 137–41
        light duty 133; installation 137
    on protected well *112*
    repair 124
    in well *132*
    working of 114
block, hardwood 103, 206–7
block and tackle 186
borehole *165*, *170*, *183*
brick flu 306, *306*
brick lining 27
bucket 97, *97*
    for excavation *31*
    foundation 80
    handmade 70
    and leading edge 99
    lidded 244
    loss of 104
    maintenance 102
    for upgraded shallow well 39
    wear in 72

and windlass 63
bucket access hole *177*
Bucket Pump 62–3, 67–8, 69–108, *71*, 253
   characteristics 72
   components
      chain 99
      concrete seal 79, *79*
      foundation 80
      headworks 104
      nuts and bolts 103
      run-off channel 81
      soakaway *82*, 83
      tools 100, *100*
      valve *69*
   fitting 76–95
   handmade 96, *99*
   maintenance 98–9
   and hygiene 73, 75
   problems 104–6
   rate of wear 72
   reverse flow 87
   and tubewell 72–4, 76, 80, 84, *85*, *86*
   village maintenance *70*, 98, 107
bucket stand *81*, 91, *92*
bucket valve 99, 100–1, *101*
   repairs 102, *102*
builders, for Blair Latrines 352
burnt bricks, for well lining 26
Bush Pump 65, *65*, 67–8, 153–209, 253
   in borehole *165*
   components
      bucket access hole *177*
      cattle trough 173
      cylinder and footvalve *158*, *164*, 185
      'down the hole' components 182
      extractable valves 192–5, 207, 209
      flexible rising mains *198*, 209
      floating washer 167
      foot valve *158*
      head assembly *162*, 199
      headworks 171, *172*, *174*, *176*
      non-extractable components 182–3
      pumphead 155, 160, 167, 184
      rising mains *163*, 195–7, *198*
      rods *163*
      seals 202
      sleeve pipe 201, 207
      soakaway 175
      tools *181*

      U bracket 190
      washing slab *175*
   fitting 180–2, 184–92
   and heavy use 167–8
   installation of 169–71
   on lined borehole 170
   maintenance 197, 199
   and protection of quality 252
   repair frequency *208*
   research 206–9
   tubewells *166*
   types *159*
      Model 'B' 160, *167*
      Model 'A' *153*, *161*
      old standard *156*, *157*
   and wear 167–8
Bush Pump, 'B' Type 153–209, *200*, *201*
   characteristics 167
   compared *208*
   description 154
   detail *203*, *204*
   fitting 180–196
   installation 169
   maintaining 197–205
   research 206
   with extractable piston *205*
   with extractable valves 207
bushes, polyurethane 127

cadmium 249
capacity, of Blair Latrine 283
casing
   being lowered *90*
   for bucket pump 78
   and concrete footing 88
   PVC 89
casing slab 179, 180
cattle trough, with Bush Pump 173
chain
   for Bucket Pump 99
   maintenance 102–3, *103*
   theft of 105
   for upgraded shallow well 39
chamber lining 94
chamber slab 94
chicken wire, use in tanks 232
chlorine
   for water treatment 266
   for purifiction 255
   residual 268
   for sterilising 264
chromium 249
cistern, for flush unit 350
cobwebs, in Blair Latrine 275
collar
   for Blair Latrine 294, 304
   concrete *34*
   for double compartment latrine 315

   protective, addition 34
   of well 33
colour of water 249
community participation 106, 107–8
   in tubewell 57
components, non-extractable for Bush Pump 182–3
concrete rings 28, 31
concrete seal, for Bucket Pump 79, *79*
containers, for sand filters 257–259
contaminants 18, 43, 44
coverplate 283
coverslab 33, 34, *37*, *175*, 177–9, *325*
   for Blair Latrine 294
   for double compartment latrine 315, 317
   for multicompartment latrine 324–5
   for shallow wells 21
currents, of air 271–3
cyanide free 249
cylinder
   for Blair Pump 123, 135
   for Bush Pump *158*, *164*, 185
   of heavy duty Blair Pump 139
   of Nsimbi 147
cylinder pipe *119*
cylinder/footvalve 150

DDF (District Development Fund) 197, 199
deep wells 21–22, *22*
Demographic Socio-Economic Study (1983) 67
desludging, of Blair Latrine 285
direct action heavy duty pump 150–2, *151*
direct action pump 142
   making 148–50
direct action reciprocating pumps 63
disposal system, waste water 237
District Development Fund *see* DDF
Dongo 217, 218
door, for pit latrine 289
double compartment latrine 312–19, 321
drainage area 91, 93
   and bananas 104
drilling *50*, *51*, 53
   and location 52
   tests 49
   tubewell 49–50
drilling rig, maintenance 56
drinking water, quality 248–9

355

drum, as filter 257–9

enteric disease 45
escherishia coli 18, 42, 43, 248, 253, 254
    counts 252
Ethiopia BP50 pump 63
evapotransporation 239
excavation, for shallow well 24, 25, 87
extractable valve system *193*
eye, of spring 215

family wells 45
ferrocement 302
ferrocement spiral Blair Latrine 309–10, *312*
ferrocement tank, moulded *234*
filter, rate of flow 258
filter
    gravel 261, 263, 265
    sand 226, 256–265
flies, in pit latrines 287
floating washers 189, 201
    system 167
floor
    for Blair Latrine 284, 297, *299*, 304
    of double compartment latrine 319
    for multicompartment latrine 326
    of pit latrine 289
    for tank and soakaway latrine 341
fluoride, in water 249
flush toilet, and Blair Latrine 349–50
flushing effect 72, 73–4
fly control, in latrines 276–7
fly trap, by vent pipe 278
flyscreen 276, 277, 300, 304
footvalve
    for Blair Pump 123, 136
    heavy duty 140
    for Bush Pump *158*
    for Nsimbi Pump 147
footvalve adaptor, for Blair Pump 126
force pumps 64
foundation, of Bucket Pump 80

gravel filter 261, 263, 265
gravel pack 57, 79, *79*, 105
gravity springs 214–15, 217
gravity well 223–4
ground water
    and bacteriological data 77
    tapping *14*
guide pipes 201

hand pump 61–8, *66*
hand washing 244, 246
    see also hygiene; mukombe; soap
handle 127, 135, 139
handle guide 123, 127, 147
handle/pushrod 150
handpump 61
    for family wells 45
    and maintenance 61
    PVC bodied 68
    reciprocating 61
    settings for *68*
    for shallow well 23
harvesters, artificial, for rainwater 226–8
    contamination in 225
head assembly 127, 199
head block
    for Blair Pump *114*, 116, *116*
    for Bucket Pump 89
    making of *91*
    mounting 117, *117*
head bolts, main, on Bush Pump 199, 201
headworks
    of Bucket Pump 104
    of Bush Pump 171, *172*, *174*, *176*
    and hygiene 43
    improving 45
health education 45
hessian 229, 230, 231
hydrological cycles *10*
hygiene 244–247
    and Bucket Pump 75
    personal 249
    at well head 43
    see also hand washing; mukombe; soap
hygienic seal 19

impermo 229
infiltration area 215
infiltration gallery 260
iron, in water 249

latrine pit, unlined 287
latrines
    directions of opening 275
    and flies 270
    and fly output 277
    and hygiene 244
    and odour 270
    orientation 276
    school 323; and cleansing 327
    ventilated improved pit 270
lead 249
lever acting reciprocating pump 65

lift pumps 64, 154–5
    mechanism 168–9
liner mould *28*, *32*
lining, of well 29–30, 59, 87
    with burnt bricks 26
    concrete 24–6
    for drilled hole 56–7
    mould for 28
    with prefabricated concrete rings 28
    of shallow wells 21
    in situ 31
    with stones or rocks 26
lizards, and Blair Latrines 281
location for drilling 52

Madzi pump 63
maintenance
    of Blair Latrine 286
    of Blair Pump 121–2, *122*, 128–30
        heavy duty 140–1
    of Bucket Pump 70
        handmade 98–9
    of Bush Pump 197, 199
    by community 108
    of drilling rig 56
    of handpump 61
    of Nsimbi Pump 146–8
manganese 249
Manicaland spring syphonage system 210, 217–19
Mark V pump 63
Matabeleland 22, 306
mercury 249
mosquitos 282, 342
moulds 28, 30, 82
mukombe *245*, 245–7, 284
Mukute trees, and soil type 24
multicompartment latrine 321, 323, 324–5, *326*, 328
Muonde trees, and soil type 24
Murgatroyd, Tommy 153

National Action Committee 67, 69
National Socio-Economic Study (1984) 67
Nira AF 85 pump 63
nitrate, in water 249, 254
nitrite, in water 249
Nsimbi Pump 63, *64*, 68, 142–152, *142*
    characteristics 143
    cylinder 147
    description 142
    detailed drawings *145*
    foot valve 147
    installation 143
    light duty *149*
    maintenance 146–8
    Mark II *144*

pumphead 147
pushrod 147–8
nuts and bolts 103, 199

outcrops, rock dome 233
overburden 14

pedestal, in Blair Latrine 297
permeability, of soil 10–11, 12–13
pipes, for Blair Latrine 275
piston valve 123, 135, 139, 147
pit
  for Blair Latrine 293, 304
  for double compartment latrine 314–15
  for multicompartment latrine 324
  for spiral latrine 309
pit latrine
  and flies 287
  and light 289
  maintenance 289
  standard 287
  upgraded 287–290, 290
  and vent pipe 287, 288
pit ventilation 271
plumb line 49
plunger 148–9
polypipe, in siphon well 222
porosity, of soil 12
press tap mechanism 235
psychodas 342
pulltite 186
pump
  see under specific pump
pump distribution 107
pump head
  of Blair Pump 135
    heavy duty 139
  of Bush Pump 155, 160, 167
    clamping 184
  for direct action pump 148, 150
  of Nsimbi Pump 147
pump head assembly *162*
pump head base plate support 126–7
pump maintenance 107
Pump minders 199
pump rods, steel 195–7
pump stand 96, 105
pumping gear 60
pump, motorised 65
purification, of water 255
pushrod 110, *119*
  for Blair Pump 123, 135
    assembly, for heavy duty 139
    lowering *121*

  of Nsimbi Pump 147–8
  repair 125, *125*
pushrod adaptor 147, 150
pushrod iron 127–8
PVC cement 118, 126

qanats 210
quality, of water 17–19, 42–4, 248–9

rainfall/water table 15, *15*, *16*, 17
rainwater
  harvesting 225–34
  storage 228
reed and grass, for Blair Latrine 307–8
reservoirs 228–234
reverse flow, in Bucket Pump 87
rig
  and operators 51–2
  problems 52–4
  setting up 49
rising main
  for Bush Pump *163*, 195–7
    flexible *198*
  for heavy duty direct action pump *152*
  lowered, 186–8
  on Nsimbi Pump 147
rods, for Bush Pump *163*
  see also pushrods
roof
  for Blair Latrine 304
    round 303, *303*
    square 302, *302*
  of double compartment latrine 319
  for multicompartment latrine 326, 327
  for tank latrine 341
roof, as harvesting point 226–7
Rotary pump 65
rotor/stator system 65
Rower pump 63
run-off channel 82, 91, 93, 95
  for Bucket Pump 81
  for shallow wells 21
  for standpost 237
  for well 59–60
rural school, and water harvesting 227

sand
  as filter 256–7
  and drilling 52
  to purify water 255
sand filter 257–9, *257*, 263–5
  for community 260, *262*
  cross-section *259*
sanitary survey 58–9, 250–1
saturation zone 11

scour pipe 215
screen, stainless steel 231
seals, replacement of, in Bush Pump 202
seepage area 38–9, 60, *240*
septic tank 334–5
shaduf 62, *63*
shallow well *20*, 21, 23, 109
shower room, and latrine 306–7, 334, 335
shutters, steel 31
single tank system 266, 268
siphon well 221–2, *222*
siphonage pipe 218, 219
siphonage system, for spring 217–19
siphonic action 221
siting
  of Blair Latrine 304
  of double compartment latrine 314
  of shallow well 23
  of well 58–9, 76
slab, rectangular 242, *242*
slab, round 241
sleeve pipe, for Bush Pump 201, 207
sludge
  rate of accumulation 335
  removal 285, 336
smoke test 272, 273, 284–5
soakaway
  and bananas 60
  brick and gravel *238*
  for Bucket Pump *82*, 83
  for Bush Pump 175
  for flush toilet 350
  for standpost 237
  with stones *238*
  for well 60
soakaway pipe, PVC 339
soakaway trench 238
soap 247, 249
  see also hand washing; hygiene; mukombe
sodium, in water 249
soil
  collapse of 24
  permeability of 10–11, 12–13
solar radiation 255, 271–3
  and Blair Latrine 275
solids, total dissolved 249
spare parts *124*, 202, 206
  for drilling rig 56
spiders, and Blair Latrines 281
spiral Blair Latrine, ferrocement 309–310, 312
spiral structure, for Blair Latrines 278, 297
sprags 178, 179

spring box 215
spring
    artesian depression 214
    and contamination 213
    and folklaw 213
    gravity 214–15, 217
    naturally occurring 210, 212–13
    protection of 213–14, *216*, *220*
    siphonage system 217
standpost 210, 236–7
steel head 123, 126
sterilising effect, of sunlight 43
stones, for well lining 26
streptococci, faecal 42, 43
subsidy, for family wells 45
sulphate, in water 249
sump 39
    base *241*
    method 239, 241
sump box 242
sunlight, sterilising effect of 43
superstructure
    of double compartment 317–18
    for multicompartment 325
    for spiral latrine 310, 312
    tank and soakaway 340
swimming pool test kit 268

tank
    brick 228–9
    cross-section *228*
    ferrocement 229, 230–1, *233*
    reservoir settling 260–1
    water storage 264
tank access hole 338
    and cover 345
tank flies 342
tank and soakaway construction 345
tank and soakaway latrine 333–6, 337–42, *344*, *348*
tap, at water point 235
Tara Pump 63
tools
    for Bucket Pump 100, *100*
    for Bush Pump *181*
training, programmes for Blair Latrines 351, 352
tripod 49
tubewell
    with Bucket Pump 72–4, 76, 85, 86
    with Bush Pump 166
    and community participation 57
    drilling 49–50, 78, *78*
    and fitting of Bucket Pump 80, 84
    and fouling 105
    hand drilled 46, 130–1
tubewell head, steel 97
tubular brick lined pits 323
turbidity, of water 249
turbulence avoidance 261
twin tank system 267, 268

U bracket, on Bush Pump 190
underdrain system, for filter 258
UNICEF 233
upgraded well 35, *39*, 253
    bacterial samples for 42
    brick lined *41*
    building 23–45
    concrete ring lined *41*
    cover 33
    excavating 24
    finished *40*
    implementation of 44–5
    lining 26–32
    siting 23
    use of 39–40
urinal
    in multicompartment unit 329, *330*
    school 329–332, *332*
    separate 331, 333

VIP 270
valve extractor, lower 136
valve replacement 126, 199
vegetable garden, and use of drainage 93
    see also seepage area
vent pipe
    as fly trap 278, *278*
    for Blair Latrine 285, 300, *300*, *301*, 304
    for multicompartment latrine 326
    and pit latrine 287, 288
    for tank and soakaway latrine 333–6, 341
ventilation efficiency 274–5
vlei areas 106
Vonder Rig 46, 48
    advantages 48
    technique 48–9, 51–4
    working parts 47

walls, of Blair Latrine 304
washing line 242
washing slab *175*, 241–3, *243*
washing stand 173
washroom, in Blair Latrine 283–4
waste stabilisation pond 335
waste water
    and bananas 239
    disposal system 237
water
    availability 10
    consumption rates 226
    distribution *11*
    infiltration of *12*
    loss of 17
    poor quality, factors 18–19
    purifying 255
    sources *44*
    storage 244
    underground 10
water-borne diseases 266
water collector 261
water consumption rates 226
water delivery unit
    see standpost
water delivery system 235, *236*
water depth measurement 183
water discharge unit 75
water diviner 24
water harvesting 225, 227, *227*
water point 48–9, 60, 251
    design 235–243
water purification 255
water quality 250–4, *249*
water supply, piped 349
water table by gravity 210
    perched *13*, 14
    and rainfall 15, *15*, *16*, 17
    rising, and quality of water 17–19
water treatment 266
Waterloo pump 63
Watt, Dr Jim 245
Wavin pump 63
weldmesh 229, 230, 231
well
    deep 21
    with Bush Pump *166*
    contaminated 20–1
    gravity 223–4
    hand-dug 14, 20
    lined with precast concrete rings 30
    poorly protected 253
    sanitary survey of 58–60
    shallow 20, *20*
    upgraded 21
    Siphon 221–2
    upgraded 35
    wide diameter 74, 84, 87, 93–5
well siting 105–6
well slab, casting 87
wide diameter well 84, 87, 93–5
windlass *21*, 36–7, 96, *96*
    and bucket 63
    on shallow wells 21
windlass supports 36, *36*

Zimbabwe Bush pump 'B' *154*